T0228412

Irrigation Horticulture
in Highland Guatemala

Westview Replica Editions

The concept of Westview Replica Editions is a response to the continuing crisis in academic and informational publishing. Library budgets for books have been severely curtailed. Ever larger portions of general library budgets are being diverted from the purchase of books and used for data banks, computers, micromedia, and other methods of information retrieval. Interlibrary loan structures further reduce the edition sizes required to satisfy the needs of the scholarly community. Economic pressures on the university presses and the few private scholarly publishing companies have severely limited the capacity of the industry to properly serve the academic and research communities. As a result, many manuscripts dealing with important subjects, often representing the highest level of scholarship, are no longer economically viable publishing projects--or, if accepted for publication, are typically subject to lead times ranging from one to three years.

Westview Replica Editions are our practical solution to the problem. We accept a manuscript in camera-ready form, typed according to our specifications, and move it immediately into the production process. As always, the selection criteria include the importance of the subject, the work's contribution to scholarship, and its insight, originality of thought, and excellence of exposition. The responsibility for editing and proofreading lies with the author or sponsoring institution. We prepare chapter headings and display pages, file for copyright, and obtain Library of Congress Cataloging in Publication Data. A detailed manual contains simple instructions for preparing the final typescript, and our editorial staff is always available to answer questions.

The end result is a book printed on acid-free paper and bound in sturdy library-quality soft covers. We manufacture these books ourselves using equipment that does not require a lengthy make-ready process and that allows us to publish first editions of 300 to 600 copies and to reprint even smaller quantities as needed. Thus, we can produce Replica Editions quickly and can keep even very specialized books in print as long as there is a demand for them.

About the Book and Author

Irrigation Horticulture in Highland Guatemala:
The Tablón System of Panajachel
Kent Mathewson

New evidence that the ancient Mayas practiced intensive, often irrigated, agriculture on a massive scale has forced revision in current thinking about that civilization. Yet, little study has focused on the heirs of this agricultural tradition; in areas of highland Guatemala, Mayan farmers today carry on forms of intensive, irrigated horticulture that suggest continuity with pre-Colombian practices.

This book examines the tablón system, a type of irrigated, raised-bed horticulture found in the present-day village of Panajachel in Guatemala. The author demonstrates how individual Mayan farmers use the tablón system as a strategy for adapting to the demands of a local economy increasingly shaped by forces of the national and international marketplaces.

Basing his analysis on a microgeographic study of tablón production and its interaction with the area's ecology and economy, he emphasizes the importance of understanding the origins and evolution of the tablón system. Mathewson points to the opportunities his approach allows for making pan-tropical comparisons with similar traditional systems of farming, and how this knowledge can be used to help revive or even introduce small-scale, labor intensive farming systems in a variety of tropical environments.

Kent Mathewson is currently a Ph.D. candidate with the Department of Geography at the University of Wisconsin, Madison. Previous appointments have included teaching at Virginia Commonwealth University, the Instituto Geográfico Militar, Quito, and the Escuela Politécnica del Litoral, Guayaquil, Ecuador.

To Lilabet for being
      there then.
To the Home Folks
      for always being here.

Irrigation Horticulture
in Highland Guatemala
The Tablón System of Panajachel

Kent Mathewson

Routledge
Taylor & Francis Group

LONDON AND NEW YORK

First published 1984 by Westview Press, Inc.

Published 2018 by Routledge
52 Vanderbilt Avenue, New York, NY 10017
2 Park Square, Milton Park, Abingdon, Oxon OX14 4RN

Routledge is an imprint of the Taylor & Francis Group, an informa business

Copyright © 1984 Taylor & Francis

All rights reserved. No part of this book may be reprinted or reproduced or utilised in any form or by any electronic, mechanical, or other means, now known or hereafter invented, including photocopying and recording, or in any information storage or retrieval system, without permission in writing from the publishers.

Notice:
Product or corporate names may be trademarks or registered trademarks, and are used only for identification and explanation without intent to infringe.

Library of Congress Cataloging in Publication Data
Mathewson, Kent, 1946-
    Irrigation horticulture in highland Guatemala.
    (A Westview replica edition)
    Bibliography: p.
    1. Mayas--Agriculture. 2. Mayas--Irrigation. 3. Irrigation farming--Guatemala--Panajachel (Sololá) 4. Horticulture--Guatemala--History.
5. Indians of Central America--Guatemala--Panajachel (Sololá)--Agriculture.
6. Indians of Central America--Guatemala--Panajachel (Sololá)--Irrigation.
7. Panajachel (Sololá, Guatemala)--Economic conditions. I. Title.
F1435.3.A37M37 1983   338.1'097281   83-17043
ISBN 13: 978-0-367-02036-1 (hbk)
ISBN 13: 978-0-367-17023-3 (pbk)

The world of Panajachel is indeed a fairy-tale world full of spirits, in which animals talk and almost anything is possible. But also it is a matter-of-fact world in which people do what they need to do. Thus other people can or must go on pilgrimages, but Saint Francis, the founder and patron of Panajachel, long ago ordered that Panajacheleños need not go, since they grow vegetables which have to be watered every third day.

Sol Tax

Contents

Tables and Maps

Photographs

Foreword

I had read Penny Capitalism, Sol Tax's classic, and various other works dealing with agriculture in the Lake Atitlán Basin of Guatemala. But nothing could have prepared me for the first glimpse that Kent and I shared of the tablones of Sololá. Spilling down the steep mountainside toward the water below, stair-stepped terraces with tablones atop each, emerged like pieces in a puzzle amidst the cloud bank that hugged the slope. Magnificent! Two days of snooping showed us the remarkable varieties of tablones in the area and the apparent intensity with which they were cultivated. Bets were quickly made as to their antiquity, knowing fully well that excavating agricultural features currently in use would be highly problematical and that dating such features would be precarious at best.

This issue aside, here was a traditional form of agriculture that in every respect, save perhaps for its areal extent, rivaled the terrace systems of the Andes, the paddys of Ifugao, and the chinampas of Mexico. And yet, to our knowledge, detailed studies of tablón construction, morphology, and cultivation was sparse. Kent Flannery's rule that the pre-eminent Latin Americanist scholars always choose some lovely highland (temperate) setting in which to conduct their field work somewhat enigmatically could not be applied to the tablón system and setting, as such. This paucity of attention was perplexing not only because of the beauty of the tablón itself—at least to the eyes of those immersed in the study of traditional agriculture—but because of its setting. The Atitlán Basin may be the most physically attractive lake basin in the world.

Tablón literally means plank or large, thick board. The term well describes the appearance of the earthen, rectangular plat-forms to which the Maya of the Basin apply it. This planting sur-face is the outcome of intensive soil preparation and is related to the generic class of activities referred to as mounding or field raising. In the Atitlán Basin, the tablón is generally irrigated and commonly terraced. As such, tablón agriculture is associated with the three principal techniques of land manipulation utilized by traditional farmers for intensive cultivation: mounding, irrigation, and terracing.

Kent Mathewson's venture into the subject centers on the tablones of the Panajachel delta, where terracing is not a particularly important element of the system. There he examined over 1,000 individual tablones, mapping the features and detailing the cultivars and cultivation practices used on them. His data provide excellent comparative materials to those described some 30 years earlier by Tax for Panajachel and neighboring towns in the Atitlán Basin. Moreover, they provide one of the few detail descriptions of tablón construction and maintenance, and the array of crops used, subjects that have implications to themes on agricultural intensification or change.

Placing this evidence in its geographical and historical context, Mathewson concludes that tablón technology was undoubtedly known in pre-Columbian times and probably used in the Atitlán Basin. Subsequent changes in the economy have changed the mode of production from a subsistence to a commodity orientation. The tablones of Panajachel are today multicropped primarily centering on vegetables for sale in markets; the resulting complex of cultivars reflects this shift in production goals. Land and market pressures have left tablón farmers with little options other than to intensify production in space and time. This intensification is achieved by careful preparation and care of the tablón and selection of cultivars that allow maximum "packing" on them, both horizontally and vertically. Increased intensification of tablón cultivation has resulted in a process of agricultural involution similar to that described by Clifford Geertz for Indonesia. Increasing inputs seem to squeeze a little bit more output from the system, but apparently at the cost of decreasing productivity and reward to the farmer.

A word must be given to the conditions in which this study was undertaken, particularly for those who may wonder why more detailed economic evidence is not provided in the assessments of the intensification and involution processes. First, as Mathewson notes, the primary focus of the study was on the identification of the crop complex and field design. Second, the field study was undertaken by Mathewson and one assistant, limiting the breadth of data that could be addressed. Despite these constraints and minimal financial support, the results of Mathewson's efforts put to shame those of some highly funded projects. In this day of big grants and lucrative fellowships, it is refreshing to be reminded of the achievements that can be obtained by a dedicated individual.

But to anyone with a touch of romanticism, the physical beauty of the tablón supercedes the discussion of data and analysis. The Basin and tablones are an inspiring sight, and perhaps a vanishing one. No descriptions of crop scheduling, no theoretical framework, and no paradigmatic shift can evoke their beauty or the feeling that encompassed both of us on that cool day in January of 1974 when we viewed that landscape. I lack the pen to capture that feeling, but I have no lack of envy for the time that Kent spent walking among those fields.

B. L. Turner II
Director
Graduate School of Geography
Clark University

Preface

Field work for this study was undertaken nearly a decade ago. It was a time of a relative peace in Guatemala's protracted civil war that now threatens to enter its fourth decade. This field work was also undertaken at the outset of an exciting decade in Maya research that has seen a major reordering of our understanding of the agricultural base of ancient Maya civilization. There is a temptation to update the material presented in this volume to take account of changes that have occurred in both the realm of Maya studies and in the world-at-large; particularly as the latter has effected the world of tablón farming in highland Guatemala. In an area of studies where the two recognized "classics"--Sol Tax's Penny Capitalism (1953) and F. W. McBryde's Cultural and Historical Geography of Southwest Guatemala (1947)--were in gestation for some ten years before publication, perhaps there is precedent for further "fallowing." There is also the pressure to enlarge this slim volume enough to give it a measure of relief against the almost monumental outlines of these earlier studies. But this has not proven practicable. With these and other shortcomings acknowledged, I offer the reader the text with only minor changes, as written during 1975-1976.

By definition, empirical studies of traditional farming systems are partially "obsolescent proof." But change does occur in even the most traditional of cultural systems. It has certainly occurred in the tablón system. I hope that the theoretical construct I have employed, viz. a traditional cultural/historical geographic outlook informed and somewhat modified by my initial readings in "World-Systems" theory, offers a durable enough basis that future workers find both the data and their interpretation useful.

Since the book was first researched, important material has appeared on: (1) the historical demography of highland Guatemala (Veblen and others); (2) the applicability of central place theory and the world-systems perspective for understanding the contemporary political/economic geography of highland Guatemala (Smith); and (3) a series of carefully done empirical studies of traditional intensive agriculture in Mesoamerica (Wilken). The bibliography has been extended to include a sampling of this work.

The basic research perspective I have attempted to follow is a path that has produced much of the best work in Latin American geography since Alexander von Humboldt's initial and colossal "breakthroughs" at the very beginning of the 19th century. Von Humboldt's travels in the "equinoctial regions," besides lifting the Spanish bureaucratic veil on outside scholarly investigation of Hispanic America, and placing the discipline of geography solidly on its modern foundations, brought about the birth of an approach that has recently been termed "Social Empiricism" (Bowen 1981:265-275). The "Social Empirical" outlook recognizes that the scientific researcher must be conscious of the social and historical context from which the empirical research is initiated, carried out, communicated, and sustained. Moreover, it requires that this awareness be directed not only toward confirming the evidential validity of this knowledge, but also toward the generation of theory integrated with a sense of social responsibility. Much of the work by cultural and historical geographers on Latin America has implicitly followed this course by focusing on themes involving traditional and tribal peoples and their livelihood systems, human modification of the region's natural landscapes, and particularly the role of state-level societies and domineering cultures in bringing about the "destructive exploitation" of local peoples and environments. Geographers working in the tradition of the "Berkeley School" of cultural-historical geography incontestably represent the core of this activism. Workers in many allied disciplines have made correspondingly valuable contributions to these kinds of studies.

It is with this sense of antecedents then, that I offer these data and interpretations of an aspect of Maya intensive farming systems in the hope that the tradition of studying such traditions is not only continued, but strengthened and expanded. It is also my hope that one day the various peoples that comprise modern Guatemala will be able to realize collectively the same measure of tranquility and humaneness that its autochthonous farmers have long imparted to their highly distinctive and humanized landscapes.

Kent Mathewson
Richmond, Virginia
July 1, 1983

Acknowledgments

While not quite capricious, acknowledging aid is an exercise that is too selective. Failure to mention the help of the "other" field workers of the area: Panajachel's horticulturalists, local residents, operators of public transport, and so on, would be a serious omission. Their composite and ill-defined aid is strongly acknowledged.

Intellectual debts are easier to trace. When I was an undergraduate, A. David Hill introduced me to the study of traditional agriculture in the Maya Highlands. He also introduced me to cultural geography and to the work of Carl O. Sauer. During the period of preparation for this study, its execution, and subsequently, Sauer's writings have been a "staple" of intellectual sustenance and a frequent tonic for provoking the imagination.

At The University of Wisconsin, William M. Denevan's advice has been heuristically imparted, and I hope well acted upon. His own work on similar problems and knowledge of various literatures encouraged me to proceed less hastily but more attentively than might have otherwise been the case. Also at Wisconsin, fellow student B. L. Turner, II, played an important role in the conception of the research objectives. He generously provided me with the means for an initial reconnaissance of the area. Professors Daniel Gómez-Ibáñez and Thomas Vale provided ancillary inspiration through their commitments to regional field studies, biogeography and conservation.

F. W. McBryde's (1933, 1947) studies served as major bench marks for my work. Until T. T. Veblen's (1975) work in Totonicapan, McBryde was the only geographer of the "Berkeley School" to focus primarily on Guatemala. Other influences came mainly from anthropologists, including several I crossed paths with while living in Panajachel. I am particularly indebted to the aid and ambience that Betty and Richard N. Adams offered at the outset. Similarly, Clyde M. Woods and his students were helpful. David Libby, U.C.L.A. graduate student in archeology and resident of Panajachel, introduced me to the geography of sacred places within the Basin. This raised a number of questions concerning the relationship between horticulture and sacred practices that forced me to look beyond (but not to investigate) the mundane limits I set for the study. During a brief visit to Panajachel, Walter Goldschmidt kindly encouraged me to design the study for a wide rather than narrow readership. In addition, L. C. Stuart, zoologist and biogeographer, was a cordial and instructive host to my periodic visits.

During the month of November (1974) some twenty anthropologists gathered in Panajachel to honor Sol Tax. A description of this symposium is recounted in Stanley (1975). At the invitation of the organizer, Robert Hinshaw, I had the chance to mingle with the participants.

In Guatemala City several individuals aided my research efforts. Sr. Flavio Rojas Lima at the Seminario de Integración Social Guatemalteca was helpful in obtaining publications of the Seminario. Dr. Francis Gall of the Instituto Geográfico Nacional was also helpful, as were the staff of the Maps and Aerial-photography Division of that institution.

Throughout the study period, Elizabeth Kennon Williams was a cheerful "compañera de campo," assisting with botanical collections and identifications and making wildlife observations. At various stages of preparation of the manuscript James J. Parsons, Hugh H. Iltis, and Miles F. Johnson made helpful comments, particularly on questions of botanical nomenclature. Mark Schultz made a trip to various tablón centers for me in early 1975 to collect comparative data. Sharon Day typed the initial drafts. Gene Dunaway, Supervisor of the School of Humanities and Sciences Word Processing Department, Virginia Commonwealth University, Richmond, Virginia, typed the final draft. I am indebted to all for their aid.

Kent Mathewson

Introduction

TABLON CULTURE

This study examines aspects of "tablón culture," a term introduced by McBryde (1947:120) to describe the system of specialized horticulture that occurs in various parts of highland Guatemala. Tablón culture is the predominant economic specialization of some Guatemalan villages, particularly in the Lake Atitlán region. Within that region, vegetable gardening for subsistence ends as well as for market trade is most intensively and widely practiced in the Sololá/Panajachel district (Map 1).

The Sololá/Panajachel horticultural area encompasses a variety of environmental zones. A continuum of tablones and micro-environments runs from the lake shore at Panajachel (1,562 m) to terraced hill-slopes (above 2,700 m) less than fifteen km away. Changes in crop composition within tablones at these different elevations and micro-environments are quite evident. The present study is focused primarily on the crop composition found within the Panajachel River delta area. From these data it seems possible to make tentative generalizations about changes within the system, particularly as they involve its expansion and intensification.

Tablones are carefully constructed raised-garden plots, generally rectangular in shape, and resemble the base of a pyramid. The sides slope at an angle of 60-70 degrees. In Panajachel they vary in height from 20 to 65 cm. They are separated by irrigation trenches formed by the tablón walls. As Wilken (1971:435) notes, "Tablón plots are carefully tended, watered, and fertilized, and production is accordingly high...."

A distinctive feature of the tablones in Panajachel are small rims, or "ears" as they are called in Cakchiquel, that are 2-3 cm high and rim the top edges of the tablón. They serve as miniature dikes to hold water on the tablón tops during the splash irrigation process.

Previous study of tablón culture has been uneven. Accounts of intensive horticulture in this area can be found in the colonial records dating from the sixteenth century. More recently, late nineteenth century travelers like Brigham (1887) or the Maudslays (1899) offered descriptions of gardening and fruit raising in the Atitlán Basin. McBryde (1933) was the first cultural geographer to discuss tablón culture. However, McBryde's (1947) massive monograph dealt with southwest Guatemala as a unit. As a consequence, tablón culture did not receive attention beyond its general role in the total economic picture of the region. Tax (1952) produced an equally ambitious monograph on the peasant economy of Panajachel. Tax describes many aspects of tablón culture, but his primary attention was given to a functional description of the overall economic activities of Panajacheleños, not just the horticulturalists. Hinshaw (1975) has recently written a "revisited" study of Panajachel. He uses Tax's work as a benchmark and offers helpful new data on the town's agriculture. None of these studies examine tablón culture as a discrete system of horticulture. Nor do they attempt to trace the origins of this system, its internal functioning, or its relationship to other village economic specializations within the Atitlán area and beyond. Therefore, this study attempts to describe the key elements of tablón culture as they function within the tablón system. In addition, the question of tablón origins is discussed. Finally, tablón culture is examined in the context of economic specialization within the region, and more broadly, how tablón culture serves as a nexus between the "two Guatemalas."

There are two distinct realms in Guatemala. The highlands (above 1,500 m) is predominantly Indian, and the agriculture is generally intensive and normally for subsistence only. This realm may be called "Inner" Guatemala. Land below 1,500 m is populated mostly by "ladinos,"[1] a Guatemalan term which refers to non-Indians in a cultural sense. In this lowland realm, or "Outer" Guatemala, most agriculture is related to the production of commodity crops for the extra-national markets. Tablón culture incorporates aspects of both realms. It is strictly an Indian activity and employs techniques of specialized horticulture and hydraulic cultivation (both labor intensive forms of cultivation). However, many vegetables grown in tablones reach the national or extra-national markets. In microcosm, tablón culture mimics the dynamics of large-scale production of commodity crops for the world market.

RESEARCH METHODS

Field work for this study was done between late August and early December of 1974. During this time we lived in Panajachel, and except for occasional trips to neighboring villages or

Guatemala City and the western Coastal Plain, I made daily observations of a variety of aspects of the tablón farming within the
town and the surrounding hills. During this period I also made
two intensive surveys of crop types and techniques. The first
survey was completed in conjunction with efforts to map the town
and locate all tablones within the delta. In the preliminary survey, a partial count of specific crop types and their distributions
throughout the delta was also recorded. This survey was begun
September 9, 1974, and completed in late October.

The final survey was made during the period of November 28
through December 5. Some 1,300 tablones were surveyed. In the
second survey, data for each tablón and all crops growing within
each tablón were recorded. The results of this survey are re-
flected in Table A.1 and Maps B.1-B.4. The main project was
mapping the Panajachel transect. I devoted roughly twenty hours
per week to mapping during October and November. The inor-
dinate amount of time spent mapping (pacing with field boots and
hand compass) may not be easily recognized when viewing the
final product. Nevertheless, the value of field work done "on the
ground" in close association with peasants and their crops inside
the micro-mileaux that form a horticultural district should be
obvious.

COMPARATIVE DATA

In the past several decades the work done on tropical shifting cultivation or swidden systems has been impressive (Conklin
1963). By comparison, the study of traditional intensive
horticultural systems has been neglected. Most of the work on
horticultural systems has been done in New Guinea and the
Southwestern Pacific area. These studies involve "primitive" or
"proto-peasant" populations. In the New World, the so-called
"floating" gardens, or chinampas, of central Mexico have received
the most sustained attention by geographers and anthropologists
(Nuttal 1920; West and Armillas 1950; Coe 1964; Armillas 1971;
Palerm 1973). Much of this work has been directed at historical or
archeological reconstructions of past systems. Wilken's (1967)
study of drained-field agriculture in Tlaxcala, Mexico, was
perhaps the first study of a contemporary intensive peasant
horticultural system in Mesoamerica. This present study attempts
to follow Wilken's lead.

Some aspects of tablón culture appear to be unique. At the
same time, elements of tablón culture suggest affinities with other
intensive horticultural systems, past and present, found in the
New World as well as the Old World. I have examined tablón culture with the hope that from this work and future studies of similar systems, a better theoretical understanding of the nature of
agricultural intensification can be constructed.

NOTES

[1] Persons of purely European descent, mestizos, as well as Amerinds who have adopted the cultural traits (language, dress, lifestyle) of the dominant Hispanic culture, can all be referred to as ladinos (Adams 1956; Smith 1976).

1
Site

LOCATION OF STUDY AREA

The town center of Panajachel lies at 14° 44' 45" N, 91° 09' 25" W (Map 1.1). The study focused on the zones of main settlement and cultivation south from the steel bridge on National Highway 1 to the lake shore. This area has been called "the delta" by Tax (1953), and is commonly known as "the pueblo." A triangle is formed by the lake shore, the east and west ridges, and the steel bridge 2.6 km upriver from the lake. The land on the slopes of the surrounding ridges is referred to as "the monte." Together, pueblo and monte comprise "the municipio."

ALTITUDE AND ZONATION

The pueblo, or delta area of Panajachel, varies in elevation from 1564 m at waterline on the lake shore to 1,635 m at the steel bridge. The river drops some 60 m from the bridge to the lake. McBryde offers a discussion of the changes in the lake level during historical times (1947:132). Periodic fluctuations occur. At the time of the Conquest the lake was about 16 m lower than it is today. Given the precipitous drop-off of the lake on its northern shore, the amount of level crop land available as the result of a lower lake level would not have been much. However, the inlet around Santiago Atitlán is shallow in places, and the level crop land available to the ancient Tzutujil capital at Santiago was probably more extensive at times in the past than it is today. Underwater reconnaissance has demonstrated the existence of submerged house sites and possibly relic field patterns around present day Santiago (Woods 1974).

In this century the lake level has fluctuated several times. In the "phenomenonally" rainy year of 1933, the lake rose 3.3 m. Generally, the lake maintains a level at about 1,554 m. Gradual increase has been the trend in recent years. The fluctuation of the lake level has had little effect on tablón farming during the

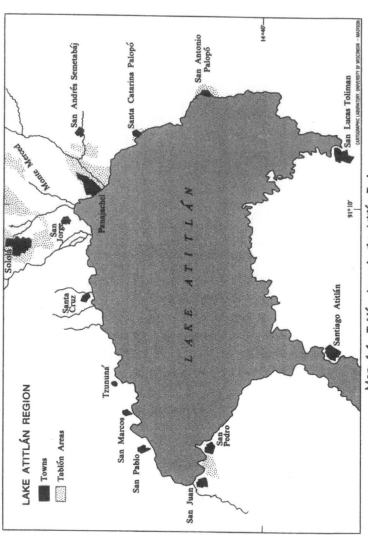

Map 1.1 Tablón Areas in the Atitlán Basin

past two decades because most of the former tablón land adjoining the lake has been sold to outsiders for building second homes. In villages with much less access to level land, as is the case in Santa Catarina, the level of the lake may play an important role in the overall productive capacities of the village. Similarly, villages drawing water from the lake for irrigation--Santa Catarina, San Antonio, and San Pedro--are, if nothing else, inconvenienced by the lowering of the lake level.

Within the municipio of Panajachel, tablones built on terraces are found on the slopes of the eastern ridge. This is a function of hydrology. The western ridge has little water. The eastern ridge receives water from streams and springs originating in the municipal territory of San Andrés Semetabáj. On the western ridge south of the Tzalá River the tablones are located between 1,700 m and 1,840 m. North of the Tzalá River (which forms a series of cataracts as it falls 300 m in a little over 500 m horizontal distance) the tablones are located on a series of artificial terraces below 1,800 m to the valley floor.

The farmers are very much aware of the microclimatic differences that occur in the relatively short 100-200 m vertical difference between the hill slopes and the valley floor. Many crops that grow well in Panajachel are not found at these higher sites. The range of microgeographical diversity is increased in remarkably small distances.

LOCATION OF TRANSECT

The area within the flood plain most intensively studied included a transect run from the lake to the steel bridge. Generally, the transect parallels the western retaining wall, which was built in the early 1950s after the particularly damaging flood of 1949 destroyed most of the tablones on the valley floor, and a section of the commercial coffee groves. The transect encompasses a section of tablones spanning 220 m at its widest width. The southwest to northeast extension of the transect encompasses a series of nearly contiguous tablón plots running slightly over 2 km.

ELEVATION OF TRANSECT

The elevational differences from the southwest end of the transect (60 m from the lake at the closest point) to the northeast terminus of tablones on the flood plain is about 50 m. The elevation at the southwest corner is roughly 1,567 m, and the last tablón in the northeast extension occurs at about 1,617 m. Despite the rise of only 50 m, the crop data shows some indication that different crop plant associations occur at the northwest neck of the transect.

4

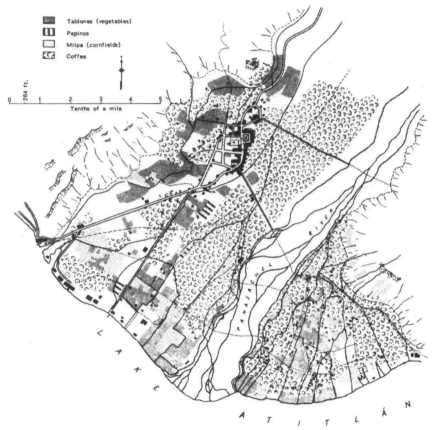

Panajachel. Irrigation ditches are shown as solid, single lines; main trails and roads, as small dash lines; old course of Panajachel River, as large dash lines.

Map 1.2. McBryde's 1936 Map of Panajachel.

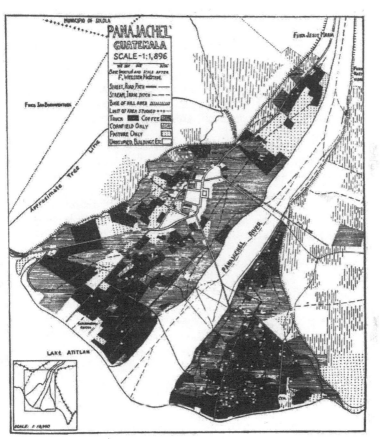

Land use.

Map 1.3. Tax's 1937 Map of Panajachel.

INCIDENCE

Tablones are prominent features of the landscape in and around Panajachel. These can be seen from vantage points above Panajachel. From the ground they are obscured by stands of vegetation, hedges, walls, cafetales, and house garden huertas. As Hinshaw (1975) has noted, the casual visitor might easily miss the well-ordered quality of the landscape. Aldous Huxley was one such tourist. He characterized Panajachel as a "squalid, uninteresting place, with a large low-class Mestizo population...." (Huxley 1934:218). Had he ventured beyond the town center, he would have noticed the predominantly Indian population engaged in their ornate gardening activities.

The tablón system around Lake Atitlán and Sololá is currently expanding (Wilken 1974: pers. com.). The antiquity of tablones in Panajachel is uncertain. Wilken (1971) has suggested that tablón agriculture is a pre-Columbian adaptation, although he has no direct evidence. Given the relatively high pre-Conquest population densities recorded for highland Guatemala, it is not unreasonable to assume that highly intensive agriculture accompanied the large population.

If McBryde's map of 1947 (Map 1.2) is an accurate indication of tablón incidence within the Panajachel delta area, then tablones have increased during the last 30 years (McBryde 1947: map 23). McBryde's map shows 60.42 acres planted in tablones.[1] Tax's (1952:339) later survey (Map 1.3) shows a much increased figure.

Tax estimated the delta land at 581 acres. Of this total acreage, Tax considered 387 acres suitable for agricultural and pastoral use. Finally, of this arable land, 142 acres were being used for "truck" gardening. In turn, truck land was divided into three categories according to crop use: corn, pepino (melon pear), and tablón. Taking this into account, we must assume a lower figure for tablón acreage per se, although land cleared for corn and pepinos not grown on tablones may represent former or future land that could be worked into tablones. Land utilized for tablones during Tax's period of observation (1936-38) would probably amount to about 100 acres throughout the delta. Field work for this study showed that tablón acreage had increased by about 10 percent since the 1930s.

The main difference today is not so much in acreage devoted to tablones, but in the shift in the areas where tablones exist. In the 1930s the main concentration of tablones was near the lake. Today the lake-front land has been sold for sites for second homes. Tablones have been moved to the more marginal areas along both banks of the river and into areas that were once cafetales but which were cleared by the flood of 1949. Tax's west section number 5 (1953:24 map 5) remains the largest contiguous expanse of tablones in the delta. It also remains chiefly under "foreign" (Jorgeño) control.

Expansion of tablones beyond the Panajachel/Sololá center has occurred during the past century. Local tradition maintains that tablones were introduced to the Almolonga/Zunil area (Quezaltenango Department) around the turn of this century (McBryde 1947:31). However, as McBryde (1947:125) also notes, "Local tradition is frequently inaccurate in this regard [dating of crop introductions]." Altee (1968: Appendix A) has estimated the size of the Almolonga/Zunil area in tablón cultivation at 2,000 manzanas. This is twice the size of the area he estimated for the Sololá/Panajachel complex. Though extent is by no means a sure indicator of antiquity, an earlier introduction of tablones to the Almolonga/Zunil complex than at the turn of this century is probably a safe assumption. One might reasonably posit a diffusion of some crops and techniques from the Sololá/Panajachel area to this higher and more westerly valley complex. Trading contact between the two areas in historic times has been considerable.

Tablón culture has diffused from the Sololá/Panajachel center to other villages within the Atitlán Basin. The best record of this process concerns Santa Catarina Palopó. Santa Catarina is the main fishing village on the lake. Since at least the 1930s there have been tablones at the lake's edge in Santa Catarina. These were owned and established by Panajacheleños (Rojas-Lima 1968:319). In this regard Santa Catarina assumes a satellite relationship to Panajachel. It continues to be a fishing village, but in recent years some of the fishermen have begun to raise vegetables. The techniques and crops were introduced by Panajacheleños.

A similar satellite relationship exists between San Pedro La Laguna and San Juan La Laguna. Early in this century Pedranos bought land in San Juan for raising vegetables. In the 1920s the Pedranos acquired more land in San Juan for the cultivation of coffee (Paul 1968:99). They got the coffee seedlings from Panajachel. At the same time the Pedranos began to introduce onions into San Pedro/San Juan tablón area. The onions probably came from Panajachel. Panajacheleños seem to be the primary innovators with many of the tablón crops that have diffused to other villages around the lake.

The corollary to spatial expansion of tablón culture is the process of "internal intensification" within the system. This is characterized by efforts to increase production through more intensive methods of cropping (especially the decrease in space between individual plants) and the new methods, such as chemical fertilization.

NOTES

[1] Rather than indicating a rapid expansion of tablones in the few years between McBryde's survey and Tax' study, it seems likely that McBryde's map of tablones is only partially complete.

2
Cultural Characteristics

CULTURAL DISTINCTIONS

The farmers of the flood plain of the Panajachel River, who are mostly Indians, fall within the categories of "traditional" or "modified" indigenes, using the typological continuum offered by Adams (1957:271-72). These farmers have become only slightly modified by Ladino influences in changing their traditional dress, customs, and worldview. The persistence of their "traditionality" and the tenacity with which they resist change has been commented upon by various observers (Tax and Hinshaw 1969).

Despite Panajachel's location as a crossroad of north/south lake traffic and access provided by a major highway running through town (formerly the Pan-American Highway), Panajachel's Indian population remains basically traditional. In this regard the town is not different from the other eleven municipios that encircle the lake.

Tablón farming is primarily an Indian occupation in Panajachel. Both Indians and Ladinos agree that tablón horticulture is an art, as well as an economic pursuit, implying that while Ladinos are strictly H. economicus, Indians are different. This conforms to Tax's earlier assessment of land use differences between Ladinos and Indians (1953:41). Ladinos tend to use their land strictly for cash cropping (i.e., coffee), while the Indians use their land for tablón horticulture. Tablón horticulture combines cash cropping with subsistence farming.

The Panajacheleño horticulturalist, like his Indian counterpart in other parts of Mesoamerica, is a participant in a "great tradition," as well as a "small tradition." Panajacheleños share cultural traits common to Indians throughout Mesoamerica. Some of these are: the corn/bean/squash triad, terracing for cultivation, and periodic markets (Kirchoff 1952). At an intermediate level between the great and small cultural construct, the Panajacheleños evince traits that demonstrate their Maya heritage. Finally, at the community level, the Indians of Panajachel possess a constellation of traits that mark them as members of a municipio that stands out from all others.

Foremost among these "patria chica" traits is an economic specialization centered around tablón farming stressing onions, garlic, and strawberries. Panajacheleños are also distinguished by their dress, speech, and system of beliefs. This does not imply that Panajachel is the only municipio in the western highlands using tablones for growing vegetables. But within the Lake Atitlán Basin, no other community specializes to the same degree as Panajachel in growing strawberries and alliaceous crops for market.

While Santa Catarina and San Antonio--the two lake towns immediately to the east of Panajachel (Map 1)--produce onions and garlic for market, they are known mainly for their other pursuits. Santa Catarina is considered a fishing village, whereas the Antoñeros are renowned cultivators of anise. Thus economic specialization becomes a defining characteristic of each municipio's "small tradition."

POPULATION

Population data for Panajachel have been recorded with some care in Tax's Penny Capitalism (1953). These figures have been revised by Hinshaw (1975). Tax's figures give a benchmark for 1936, while Hinshaw's work provides demographic information for 1964. When Tax studied Panajachel in the 1930s, the number of native Panajacheleños was 780. In addition, there were around 300 Ladinos and 100 "foreign" Indians. In 1964 the population of Indians was 2,023, including some 400 "foreign" Indians. The number of Ladinos had grown to 1,245. Thus the total population of Panajachel had grown from around 1,200 in the mid 1930s to 3,268 by the mid 1960s. The population had more than doubled just in the span of one generation.

POLITICAL ORGANIZATION

The internal political organization of the Panajacheleños is a variant of the polity found throughout indigenous Mesoamerica. Described as a "civil/religious" hierarchical form, the phenomenon has received sustained attention in the anthropological literature (Cancian 1967). Detailed accounts have been published for a variety of communities in the Guatemalan highlands. Tax's remarks on the region as a whole suggest that just as each municipio has its particular economic specialization, each municipio has evolved its own distinct civil/religious hierarchy (1937:442).

While no attempt was made to gather information on the specific nature of the civil/religious hierarchy in Panajachel, several questions arise that concern the possibility of links between the persistence of traditionalism within Panajachel and the intensification of its agriculture. Following Brookfield's (1972) admonition, any attempt at explanation of agricultural intensification within traditional societies should examine the role played by production for social or ritual ends. In the case of traditional communities in Guatemala, the cost of participation in the civil/ religious hierarchies is not expressed in terms of labor time alone (Wolf 1957). The cargo system also demands expenditures of money and for agricultural produce to service the religious fiestas. Tax (1952:58) suggested that there might be an inverse correlation between the wealth of a community and its level of commitment to sponsoring fiestas. The smaller communities may even be held at relative levels of poverty through devotion to expensive fiesta habits (Cancian 1967).

At times in Panajachel the costs have become so great that the cargo system has been weakened through refusals to serve. Similarly, some of the motivation for "foreign" Indian migration to Panajachel during this century can be directly attributed to attempts to avoid the burdens of the cargo system. Several farmers from the village of San Jorge moved to Panajachel in the 1920s to escape the system in their native village (Tax 1946). These same farmers have become established in Panajachel and are recognized as skillful gardeners. Conversations with present members of their families suggest that irrigable land in and around San Jorge has become circumscribed to the point that migration to Panajachel was seen as a necessity. Therefore, the pressures of social production in at least one case forced spatial expansion (migration) of tablón farmers into the Panajachel delta.

SOCIAL ORGANIZATION

Using Wolf's (1957:1) definition of a peasant "as an agricultural producer in effective control of land who carries on agriculture as a means of livelihood, not as a business for a profit," most Panajacheleños who work tablones for a living would have to be excluded. Some of the more "wealthy" Indian land owners employ landless and landed Indians in one or all of the stages of tablón production. In a limited sense they do extract a profit, not all of which is returned to the community (or lost) through ritual obligations. The landless Indians are at the other extreme. They work for tablón owners on a daily cash basis. Are they to be excluded from the ranks of the peasantry as well? Or can they be described as a rural proletariat serving the minifundiarios? The problem is one of scale; the answer is not clear cut.

In terms of the Indian community, the landless Indian worker is a peasant by virtue of membership in a peasant community. However, when the landless Indian or the minifundiario goes to the coast to work on a plantation, they clearly assume a proletarian status vis à vis their social relationship to the "means of production."

Rather than attempt a typology of peasants and nonpeasants within the Indian community of Panajachel, for the sake of brevity it can be suggested that the community as a whole is peasant in nature. Despite the landlessness of some of its members, participation in the traditional society is not seriously restricted for any native Panajacheleño. Therefore it seems reasonable to characterize all those involved in tablón farming as peasants.

LINGUISTIC AFFILIATIONS

The municipio of Panajachel is in the western part of the Cakchiquel language area, and Panajacheleños speak a variant of Cakchiquel. From Panajachel, the area extends southward into the Boca Costa and eastward to the old colonial capital of Antigua (Map 2.1). Three major Maya languages are spoken within the Basin. Cakchiquel is spoken by peoples of the northern and eastern shore, while Tzutujil is spoken in Santiago Atitlán and along parts of the southern shore. Quiché is spoken on the western slopes of the Basin. The linguistic variety is compounded because each municipio has its own dialect.

Early ethnographic observers in the highlands spoke of ethnographic divisions along linguistic lines. For example, all speakers of Cakchiquel were thought to be members of a single nation and to express close ethnological affinities. Much of the impetus for this early speculation probably came from the German tradition of philologically-oriented culture study. Schultze-Jena's (1945) work provides an example. He assumed that Quiché culture was a unit. If this were so, presumably one could use data from differing communities like Momostenango and Chichicastenango interchangeably.

Tax (1937) pointed out the fallacy in this approach. Given the linguistic diversity at the municipio level along with economic specialization and other aspects of material culture, the ethnic unit may be defined in terms of municipio--but not language. A cultural ecological analysis might extend this approach and in a different direction. One might look to other examples in the highlands where tablón-based specialized horticulture is practiced, particularly Almolonga and Aguacatán (Map 2.1), and search for cultural regularities between these linguistically and environmentally disparate places.

LARGER GROUP IDENTIFICATION

The tablón farmers of Panajachel are part of two larger groupings. Foremost, they are Maya Indians. To a lesser extent

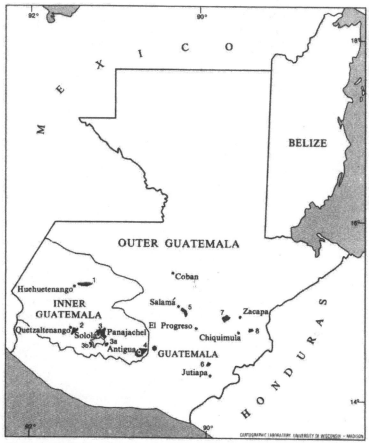

Map 2.1. Major Vegetable Producing Areas of Guatemala.

I. Inner Guatemala
   (Tablón Culture)
   1. Aguacatán/Buca River Valley
   2. Quezaltenango
      (Zunil/Almolonga)
   3. Sololá/Panajachel
      a. San Antonio Palopó
      b. San Pedro/
         San Juan Laguna
   4. Antigua Guatemala

II. Outer Guatemala
    (Ladino Modes of Production)
    5. Salama/San Jeronimo
    6. Laguna de Retana
    7. Teculután/Zacapa
    8. Chiquimula/Jocotán

they are Guatemalan nationals. In their own estimation, their indigenous identity takes precedence. The Maya are concentrated in the highlands of Chiapas, Mexico, and throughout highland Guatemala. They represent large portions of the population of the Yucatán Peninsula and occupy zones of varying extent in Belize, Honduras, and El Salvador. To a lesser extent Maya peoples have begun to reoccupy the northern and eastern lowlands of Guatemala, especially around Lake Izabál. These later immigrants are Kekchi Maya from the uplands of Alta Verapas (Adams 1965). In the Guatemalan highlands the population is predominantly Maya-speaking Indians. In many areas Indians represent over 95 percent of the population. Most non-Indians in the highlands live in towns or cities. The rural population is overwhelmingly Indian.

SUBGROUPS

It can be suggested that the cultural homogeneity within municipios (particularly within Panajachel and its surrounding hillsides) is a fairly rigid arrangement. Even though the bounds of the municipio usually delimit ethnic units, considerable immigration from other municipio occurs. Internal colonies of "foreign" Indians exist in many municipios. For a municipio the size of Panajachel (3,208 in 1964 including resident Ladinos), the number of "foreign" or non-Panajachel-born Indians is large. Hinshaw has calculated that foreign Indians represent some 30 municipios, with the majority hailing from the neighboring municipios of Sololá, San Jorge, Concepción, Santa Catarina, and San Andrés (1975).

The rigidity of the concept "puro-Panajacheleño" vs. "extranjero" (foreigner) became apparent through conversations with Indians proud of their residential purity. Even though an Indian family may have lived in Panajachel for several generations, the fact that their predecessors came from another municipio (sometimes only a few kilometers away) meant that they are forever "extranjeros" in the eyes of the Panajacheleños.

There are two major attractions for immigration. The tourist industry is the main attraction. Panajachel supports seven hotels and a number of smaller pensiones. This provides jobs in construction, maintenance, food service, and other similar occupations. Another facet of Panajachel's relative affluence is the number of second homes owned by Guatemalans, North Americans, and Europeans. These periodic vacationers (some 100 families) employ a number of Indians as caretakers, gardeners, maids, and cooks.

The second attraction, though somewhat less accessible in the wake of rising real estate prices, is Panajachel's level flood plain. It is used extensively for irrigated gardening. A colony of Jorgeños from the neighboring municipio of San Jorge has established itself throughout the western side of the town. Many

Jorgeños own land in Panajachel but keep their main residences in San Jorge. Others have established their main residences in Panajachel and keep secondary houses in San Jorge. The "second homes" are used for occasional visits, or during the periods men serve their charges within the civil/religious hierarchy. Another group of Jorgeños--day laborers for the Jorgeños who own or rent land in the delta--walk several kilometers a round trip each day from Panajachel. Their travel time is an hour or more each way because the 4 km trip entails traversing several hundred meters of extremely steep trails.

To a lesser extent, day laborers from other municipios come to work in horticulture. Another group of "hill" people make daily trips to Panajachel for wage labor in the tablón plots. They live in the dispersed settlements associated with the municipio of Concepción. Coming mainly from the settlement of Monte Merced--at above 2,000 m. (Map 1.1)--they traverse steep slopes descending to the Panajachel River (at 1,600 m.) and up the east side of the valley to tablón lands around Media Cuesta (1,700 m.) overlooking Panajachel. This is another example of day laborers expending considerable amounts of time and energy while in transit to work tablón plots.

Time did not allow information gathering on territorial domain of each group of foreign Indians within the Panajachel tablón district to be completed. However, identification of zones can be offered (Map 1.3). The Jorgeños are concentrated on the western side of the river. This area contains the majority of "foreign" Indians, Ladinos, and second-home owners. The east side of the river is distinguished as a somewhat separate barrio known as Jucanyá, the residential neighborhood or stronghold of the traditional Panajacheleños. Most of the tablones owned or rented in Jucanyá are similarly controlled by Panajacheleños.

In the northern apex of the delta (west bank of the river), the land is largely owned by Panajacheleños but rented or worked in part by Indians from the Monte Merced area or recent migrants from Sololá. The eastern ridge around Media Cuesta is almost entirely worked by Indians from Monte Merced. No firm information on the terraced tablones of the eastern slope north of Media Cuesta was gathered. However, several Panajacheleños spoke of owning tablón land in this sector. The overall picture probably represents a mosaic of ill-defined ownership patterns, rather than strictly segregated microdistricts. This generalization applies as well to residential settlement. Only Jucanyá demonstrates a noticeable concentration of Panajacheleños to the exclusion of "foreign" families.

3
Origins of
Tablón Culture

Alvarado and his brothers, together with others, have killed more than four or five million people in the fifteen or so years from 1524 to 1540 and they continue to kill or destroy those who are left. They have destroyed or devastated a kingdom [Guatemala] more than one-hundred leagues square, one of the happiest in the way of fertility and population in the world.

Bartolomé de las Casas, 1552

Brevisima Relación de la Destrucción de las Indias

EVIDENCE OF PRE-CONQUEST TABLONES

Negative Evidence

Veblen (1974:313), who suggested that tablones are not pre-Columbian in origin, based his argument on the crop composition of tablones at one center. He rejected Wilken's (1971:534-36) statement that tablones may be of "pre-Columbian ancestry because many of the plants so tended are native." Veblen found "a vast majority of the plants raised on the tablones are non-American vegetables such as onions, garlic, cabbage, lettuce, carrots, radishes, broccoli, and cauliflower."[1] He estimated that less than 20 percent of the crops were native in origin. By giving only a species count, he overlooked other factors.

Primarily, Veblen failed to differentiate between crops marketed and crops directly consumed. In Panajachel there is a very noticeable differentiation between "Ladino" crops bound for market and native crops consumed by the farmer and his family. Moreover, native crops can be less conspicuous to the Euro-American observer, especially root crops. If bulk weights of native crops are compared with the European market crops, then the disparities may not be so great.

There are other objections to Veblen's observations. First, he failed to distinguish between the environmental conditions of the Almolonga/Zunil center and the tablón areas in the Lake

Atitlán Basin near lake level. The Almolonga tablones occur at 2,300 m, and in Zunil they are above 2,000 m (McBryde 1947:32). On the flood plain of the Panajachel River they occur between 1,570 m and 1,600 m. The mean difference of 500 m between the two vegetable centers is compounded by the moderating influence of the lake Panajachel. Moreover, the tropicality of Panajachel's micro-milieux favors a broader range of native crops. In contrast, most of the popular European vegetables are best adapted to temperate or cool climatic conditions. Therefore, it is not surprising that the Almolonga/Zunil area specializes in European introductions.

Finally, the Almolonga tablones may be very recent. According to McBryde (1947:31): "This is said to be a recent center--important only since about 1910." The techniques of tablón gardening, according to local tradition, were introduced to the Almolonga area by onion seed traders from Sololá. If this dating is correct, then the predominance of non-native crops is hardly surprising.

Positive Evidence

Wilken (1971:435-36) and Turner (1974:71) have argued that tablones antedate the Spanish arrival. It is worthwhile to quote Wilken's assessment in full:

I suspect the tablón of pre-Columbian ancestry although I have no firm evidence for this. Fuentes y Guzman provides a detailed description of a tablón-like system in seventeenth century Totonicapán which involved irrigation, transplanting, and protection of crops from hail. Many of the plants so tended are native, which suggests that the whole complex of techniques may have had pre-Spanish origin. Tax notes that at Panajachel Indians tend to grow truck while Ladinos grow coffee. This may represent an old tradition although Tax remarks that it is impossible to say whether the farming techniques were introduced after the Conquest or were adapted from old Indian systems.

Unless precedents for tablón farming can be found elsewhere, it seems likely we are dealing here with another indigenous system of intensive cultivation.

Turner's opinion is more emphatic: "This author feels strongly that tablones are an aboriginal highland technique of cultivation and that investigations of these features will prove this."

While no attempt was made to date tablones by archeological methods, this author agrees with Wilken and Turner on the grounds that tablones share enough structural and functional similarities with chinampas to suggest that they are variants of chinampas, or vice versa. Chinampa techniques could have been introduced into the Guatemalan highlands at any point after the Early Classic (A.D. 600) by the successive waves of Mexican emigrants. Moreover, one could suggest that chinampa techniques were adapted to riverine flood plains in highland Guatemala (creating tablones) because the lacustrine areas of highland Guatemala are less suitable for extensive chinampa farming than similar areas in highland Mexico.[2]

Quite likely, tablones were used by highland Maya for intensive vegetable gardening in a manner not unlike they are used today. The assemblage of crops would have been different, but not radically so. Humboldt (1966:274) felt that aboriginal Americans grew onions and possibly garlic: "We know with certainty that the Americans have always known the onion (in Mexican, Xonacatl)." Therefore, even the dominant crops of the Panajachel area may have been present in pre-Columbian times. No European crops are grown at all on many tablones in Panajachel. Tablón culture is by no means dependent on European crops.

Assuming that tablón horticulture was technically and culturally within the ken of Maya farmers, we turn to other reasons for its possible origins. Perhaps the simplest explanation has been suggested by Adams (1967:85; 1974: pers. com.). He has suggested that highland agricultural intensification underwent a qualitative change through adaptations designed to exploit small flood-plains exposed to periodic flooding. This notion offers a theory of tablón genesis: rocky flood plains, otherwise unfarmable, become extremely productive by using tablones. Groups did not have to be highly organized in the Wittfogelian sense to exploit these niches. They could have built tablones on the flood plain benches with the expectation of periodic destruction of their works, but not at a reconstructive cost outweighing the long term returns. This strategy is used in Panajachel today.[3]

The highly important pre-Columbian cacao industry provides another alternative. Tablones are sometimes used for the nurture of coffee seedlings. John Vickers, English coffee planter and resident of Panajachel in the early part of this century, may have been the innovator. Using tablones for coffee seedlings may also point to a practice that could have begun with cacao seedlings in pre-Conquest times, and adapted to coffee nurture during the past century. The pre-Columbian cacao industry used irrigation and intensive cultivation (Millon 1955; Bergmann 1969). As previously mentioned, the ancient Tzutujil capital of Atitlán had extensive cacaotales. Tablones may have been used for cacao nurseries, and then brought to Panajachel before 1560 with its first post-Conquest inhabitants--a group of Tzutujil colonists from Atitlán (Lothrop 1933:103).

Similar to the cacao nursery thesis, tablones may have originated as ceremonial garden plots for growing ritual or tributary crops.[4] In this form, tablones might have been used as early as the Classic period (A.D. 300-1000). At that time the theocratic rulers from Teotihuacán (in highland Mexico) established hegemony over much of the western highlands and exacted tribute from the Maya. The role of ceremonial gardening in connection with the chinampas of the Valley of Mexico is well documented (Nuttal 1920). Today, the center of chinampas in Mexico is Xochimilco. In Nahuatl, <u>Xochimilco</u> means "place of the flower field." The origin of chinampas probably dates back to the Classic period or earlier (Armillas 1971). Therefore, chinampas or their variants--tablones--may have been introduced in Guatemala for the production of ceremonial crops, especially flowers.[5]

The extreme care and aesthetic attention given tablones by modern Maya farmers strikes many observers as going beyond the bounds of the simply profane. On this Tax (1953:129) said, "A good deal of pride is involved, and no doubt good workers keep their gardens better appearing than is technically necessary."

Further evidence implies a connection between tablones and ceremonial origins. As part of the Day of the Dead ceremonies (November 1 in Panajachel), miniature tablones are constructed on top of the grave sites of former farmers. They are made of earth carried from the fields and overlaid with pine needles spread on the top surface. This replicates the covering used by most farmers on their seedbeds. Apparently, the symbolic significance of these miniature "necro-tablones" has not been studied. Anthropologists often point to such practices as indicators of cultural continuity and antiquity.

Without further searches in the documentary sources and without archeological investigations of terracing associated with tablones, no conclusive statement on the age of tablones in the Atitlán Basin can be made. Nevertheless, arguments can be made that tablones are pre-Columbian in origin. Tablones probably did not spring from one single impetus. Rather, they probably served a variety of secular and ceremonial purposes much as they do today. The chief purpose of tablones is to provide an efficient mechanism for highly intensive vegetable production. The pressures that bring about the shift from long fallow, extensive agriculture to short fallow, highly intensive methods are partly due to population pressures, but also to those pressures Brookfield (1972) has called "the pressure of needs." These include production for trade, production for social ends, as well as production for direct consumption. The way in which this total ensemble of "need pressures" effected agricultural intensification and disintensification in the Atitlán area in post-Conquest times is discussed below.

POST-CONQUEST TABLON DEVELOPMENT

Little study has been made of the economic and agricultural history of post-Conquest Guatemala. C. L. Jones' (1940) history of Guatemala is an exception, which emphasized changes in labor arrangements. More recently MacLeod (1973) has published a general socio-economic history of Central America from 1520-1720. Drawing upon the work of Berkeley cultural geographers and historical demographers, MacLeod's study offers an appropriate framework for interpreting the colonial period. MacLeod stresses the "boom and bust" nature of plantation agriculture and other efforts in Guatemala to produce products for the world market. This perspective is helpful in understanding the history of changes in the interplay between peasant ecotypes in both Inner and Outer Guatemala. [6]

A synopsis of the main changes in the agricultural history of the highlands is possible. Major changes in the modes of agricultural production provide a convenient focus. Since the Conquest, the strategies for labor control and procurement have assumed several juridical and operational forms. In gross relief, these changes correspond with important reorientations in Guatemalan agro-commodity production. Three distinct periods can be identified:

1) 1521-1541, the Conquest with imposition of slavery and seizure of land,
2) 1542-1616, the search for " produits moteurs" and the imposition of repartimiento and encomiendas, and
3) 1616-1944, the use of mandamientos culminating in the coercive cash-cropping systems under the "vagrancy laws." [7]

Though this scheme oversimplifies Guatemalan economic history, it allows for a useful periodization of the post-Conquest era. Moreover, it allows us to identify possible sources of pressure that effected the intensification of agriculture, even in localized situations, e.g., in the Lake Atitlán Basin.

THE INITIAL IMPACT OF THE CONQUEST, 1521-1541

The first two decades of Spanish invasion and conquest established patterns of use and abuse of land and native peoples that continues to the present day. The effects on agricultural intensification, or--in many cases--disintensification, were registered in two broad categories. Native livelihood patterns were severely disrupted by depopulation, military occupation, and forced reorganization of settlement patterns. Less violently, native life was effected by cultural introductions, both Mexican and Mozarabic in origin. We are mainly concerned with how these changes, regenerative as well as disruptive, may have affected aboriginal maintenance of intensive forms of agro-production such as tablones.

POPULATION DECLINE

In many parts of the New World, the first years of contact between the Americans and Europeans were catastrophic (Denevan 1976). Highland Guatemala was particularly affected. The chaos and internicine warfare among the highland Maya tribes during the later pre-Conquest period (ca. 1400-1524) was compounded by the Spanish intrusion. The highland Maya had not reached a level of socio-political organization comparable to the Aztec imperium. In comparison, the Spaniards were unable to assume control by simply decapitating the state structure. The Spaniards had to wage a campaign of systematic societal disruption to conquer the highland Maya.

The most immediate effect of the Conquest was destruction of the native population. Depopulation was severe and took three major forms. First, the highland Indians were struck with a series of epidemics (McBryde 1940). Las Casas (Mackie 1924:132) estimated that four to five million, or one third of the population, was killed during the plague year of 1522.[8] Apparently the epidemic was spawned by the Spanish in Mexico. Aztec agents sent to warn the Maya of the new invaders probably were the unwitting vectors. Whatever the source and manner of transmission, it acted as a superb vanguard for the Spanish invasion two years later.

Second, though the Spanish accounts are often exaggerated, tens of thousands of Indians were killed in battles with the invaders. This is corroborated by the Indian's own written accounts. In addition, the highland Indians waged war among themselves, sometimes at the direct instigation of the Spanish, but more often as a continuation of pre-Conquest rivalries.

The third major cause of depopulation occurred in the wake of agricultural disruption. The destruction of native agriculture was immediate, but also protracted. Summary disruptions were caused by warfare. Pedro de Alvarado, the conquistador of Guatemala, used field burnings to intimidate the Indians. As Mackie (1924:59) reports, Alvarado brought "the Indians to terms by threatening to destroy their fields." Other methods of agricultural "destabilization" included setting livestock out to pasture in native fields and gardens.

Brookfield (1972) has pointed out that sudden depopulation often decreases the intensity with which a population farms. To survive the first several decades following the Conquest, Indians may have foraged wild foodstuffs or reverted to swidden cultivation.[9] Pressures to produce ritual or tributary crops like cacao, tobacco, and flowers were undoubtedly relaxed. During the hiatus between collapse of the native rulers and the imposition of Spanish control, tablones for official purposes could have fallen into disuse. This happened with many of the cacaotales (often irrigated) requiring specialized attention (MacLeod 1973:70).

MILITARY OCCUPATION

Military occupation affected agricultural modes of production. First, this occupation resulted in reestablishing a civil order in some ways similar to the pre-Conquest status quo. For the common laborer, tribute to hierarchical authority still had to be paid. Given the depopulation, the tribute was more difficult to meet than before.[10] Second, renewed tributary pressure may have promoted agricultural intensification. Tablones abandoned during the more chaotic moments of the Conquest would have been revitalized to meet Spanish demands for food stuffs.

Another problem for native agriculturalists was sown in the period of direct military occupation. Allowing cattle and horses to forage in native milpas began as a martial tactic but soon developed into a prolonged problem for the Indians. Gibson (1952:152) paints a picture for Central Mexico that was duplicated in Guatemala:

It is not easy for a modern reader to assess properly the grave economic effects of such [a deprivation]. For the Indian agriculturalists the damage caused by cattle was overwhelming. Animals devoured Indian crops by night and day. Spanish cattlemen permitted the animals to enter Indian areas before the specified time, and damage was particularly heavy during August and September, when crops and the steers and oxen brought from Veracruz to Mexico by the carreteros, destroyed the Indian's maize, vegetables, fruit, and tunales. The rule that cattle were to stay one-half league from Indian plantations was constantly violated. Houses were forcibly entered and ruined; public works were damaged; whole pueblos were destroyed and boundaries that had been made were overrun. In 1594 one of the largest towns of the province, Hueyotlipan, was temporarily abandoned by the Indians after the destruction of its nopal and fruit crops by roving cattle.

In the face of this livestock invasion, Indians may have been forced to adapt more intensive modes of agriculture as a defensive measure.[11] A manzana of tablones, surrounded by larger irrigation ditches, would have been easier to guard than hectares of milpa, which were often separated from settlement sites.

FORCED SETTLEMENT RELOCATIONS

The first strategy for labor control was simple slavery. This began as a result of the military campaign--the vanquished were often enslaved. By the 1540s the influence of the Church began to moderate the methods by which the Indians could be forced to labor. In order to proselytize and make the Indians produce more efficiently, the Church sponsored a broad resettlement policy known as "reducción" or "congregación." The surviving Indians were encouraged or forced to resettle in villages designed in many cases by the clergy.

The search for gold was another disruptive development that characterized the first two decades of Spanish occupation. Placer mining along stream beds demanded considerable labor. Enslaved Indians were often used; the practice of slavery as a form of labor procurement was reenforced by the search for gold, which led to population relocation. As MacLeod (1973:121) suggests: "Many of the pre-Conquest villages had been on defensive sites... quite far away from the valley 'bottoms....'" When confined to the circumscribed domain of the flood plains, Indians would have turned to more intensive methods of cultivation.

In addition, much of the former land the Indians had used for extensive agriculture was taken over by livestock, both feral and tended. Tablones would have been a logical adaptation in the face of such pressures.

TABLONES: A POST-CONQUEST MEXICAN INTRODUCTION?

It is conceivable that raised-field agriculture and irrigation techniques were introduced into the highlands along with the Spanish invasion, and not as part of the agronomic baggage imported from Mexico during the Teotihuacanian invasions (ca. A.D. 600). First, Alvarado enlisted an army of Indians from Central Mexico--chiefly from Cortes' allies among the Tlaxcalans and Mexicans--to fight the highland Maya in Guatemala. The Mexicans, and probably the Tlaxcalans, were skilled intensive farmers (Wilken 1967). They used irrigation plus drained and raised-field techniques in their domains to support large populations. They may have transmitted these techniques to the areas they colonized in Guatemala after the Conquest.

At present, many of the areas of highland Guatemala that practice intensive cultivation carry Mexican-derived place names (Arriola 1973). Aguacatán, located in the Department of Huehuetenango, is one example. At the time of the Conquest it had a dense population comprising twelve villages bearing Maya names (Recinos 1954:242). The twelve former villages were reduced to two towns--Aguacatán and Chalchitán. Both are Nahua Mexican names.

Aguacatán is one of the most important onion and garlic centers in Guatemala.[12] The techniques and tools used in raising these crops differ from tablones in the Panajachel/Sololá hearth. A more primitive adze is used to make the narrow raised garden beds.[13] The irrigation ditches are V-shaped rather than the trapezoidal shape used in Panajachel. This seems to be a function of the differences in the shape of the two styles of adzes employed at each center. In addition, the bottle gourd (Lagenaria spp.) is still used in Aguacatán for splash irrigation. On the basis of the primitiveness of the tools used in Aguacatán, these tablones may be an arcane form of raised-bed horticulture introduced by invading Mexicans at the time of the Conquest, if not before. The entire question of Mexican introductions--plants, animals, tools, and techniques--deserves more study.[14]

AN IBERIAN ORIGIN OF TABLONES?

The Spaniards had a major impact on native agriculture, chiefly through plant introductions. This process began immediately after the Spaniards arrived (Veblen 1974:336). Plant and animal introductions were part of the Church's efforts to reorder the aboriginal landscape (Brand 1951:122).[15] Iberians were acquainted with techniques not unlike the intensive forms of horticulture practiced by the Indians. Irrigation agriculture flourished in Spain several centuries prior to the Conquest of the New World. By the tenth century Islamic Spain had become a fully "hydraulic society," according to Wittfogel (1957:215). If the earliest conquistadors were products of Spain's cattle-raising western frontier, as often suggested (Foster 1960:62), they may have been unfamiliar or antagonistic to the Arabic horticultural tradition. But by the 1540s a retinue of clerics and soldiers-of-fortune from many parts of Spain began arriving in the highlands. Among these adventurers were persons capable of transmitting new crops and techniques to highland huertas.[16] Though long-distanced diffusion is not needed to explain the origins of tablón agriculture, it has undoubtedly modified the practice.

CONSOLIDATION OF THE CONQUEST: 1541-1616

The years 1541-43 were significant in a number of ways for Guatemalans. In as many years, three events happened that signaled major changes in strategies for producing commodities for the world market, settlement policy for Indians and Europeans, and labor arrangements concerning the remaining Indians.

First, Pedro de Alvarado was killed in Mexico at the outset of an abortive expedition to the Orient. He sought the gold and spices that had largely eluded him in Guatemala. His death coincided with the first efforts by the Spaniards remaining in Guatemala to develop spice and condiment production in situ.

Second, in the fall of 1541, Almolonga (Sacatepéquez), the capital of Guatemala since 1527, was destroyed by a massive flood and mud slide. Almolonga had been built at the foot of the volcano Agua. Locals blamed water that accumulated in the crater and was pushed down the mountain by an eruption. It seems more likely that a combination of eruptions, rains, and erosion from the deforested slopes created the disaster (Jones 1940:16). As a result, a new capital was built for a third time with the nearby plain of Panchoi being chosen for the site. The disaster may have also spurred attempts at regulating deforestation. In turn, this probably influenced the efforts at restricting shifting cultivation in favor of permanent-field farming.

New villages were planned around the capital to aid relocation and "reduction" of the Indians. Each village was planned as a specialized craft or cropping center. In one village the economic activity was fruit raising, in another metal smithing, and so forth. This mirrors the pattern of economic specialization that exists in parts of the highlands today, an arrangement that is particularly evident in the Lake Atitlán Basin. The question of how much this spatial ordering of production was the result of Hispanic planning, and how much it is a legacy of pre-Conquest production patterns merits more study. [17]

The third event had equivocal results. After Alvarado's death, Guatemala was nominally leaderless. During this same period, the Church attempted to expand its influence throughout the New World. Bartolome de las Casas, critic of the defacto slavery and abuses carried out under the encomienda system, persuaded the Crown to issue the famous "New Laws," which codified and reformed labor arrangements between Indians and Spaniards. Although the New Laws were protested and--in many cases--ignored, it was a victory for the Church. The New Laws enabled the clergy to involve themselves more directly in the resettlement of Indians, thereby gaining additional influence in molding the livelihood patterns of the Indians.

THE SEARCH FOR VEGETAL PRODUITS MOTEURS

After chattel enslavement of Indians was outlawed and the yields of gold and silver proved modest (MacLeod 1973:64), many "strange and ephemeral products" were turned to in the search of a cash crop. Balsam, liquididambar, canafistula, and sarsaparilla all were sought in hopes of quick profits. Each enjoyed a minor boom, but the market, as well as many of the plants, had unelastic potentials. The exploitation of these plants did little to change the modes of agricultural production in the highlands. Another cash crop did.

Cacao was an extremely important "cash" crop in pre-Conquest Mesoamerica; indeed the cacao bean was specie (Millon 1955; Bergmann 1969). Along the Pacific versant, and elsewhere, the highland Maya had many cacao plantations. The Spaniards began to turn to cacao as a cash crop after 1540. Between 1548 and 1551, Atitlán, the Tzutujil capital, collected the largest amount of tributary cacao of any town in Guatemala (Bergmann 1969:90). Much of this cacao came from Atitlán's plantations in or below the Boca Costa. The highland Maya exploited a variety of niches not necessarily contiguous with their own core area. This system was probably similar to the strategy of "verticality" employed by the Inca in Peru (Murra 1972). Pressure to expand cacao production increased during the 1550s and 1560s.

Using the encomienda and repartimiento systems, rather than slavery per se, the Spaniards demanded increasing amounts of cacao from the Indians. The burden was great enough on the Atitlañecos that their noblemen protested the terms of the cacao tribute required after 1540 (Jones 1940:116). As depopulation of the lowlands and the Boca Costa continued, more highland Indians were sent through the repartimientos to work in the coastal cacao groves. Unskilled and uprooted, these new workers proved less effective than the former growers. Blight, deforestation, and cattle encroachment all created problems for the coastal plantations. By the end of the sixteenth century, the cacao industry was in disarray. In addition, by forcing highland Indians to migrate to the lowlands and the Boca Costa, the groundwork was laid for the later labor migrations out of the highlands.

NEW TOWNS AND REDUCCIONES

Foster (1960:67) voiced a generalization about the Conquest that lacks accuracy when applied to Guatemala:

The introduction into America of Spanish agricultural practices contrasts markedly with the introduction of city planning and building, since informal and unguided processes rather than formal direction explain the selection that was made of elements from the donor culture.

This history has yet to be written for Guatemala. From a cursory look at the materials, one would have to conclude that in many cases new towns and reducciones were joint efforts in urban and agronomic planning. Many writers have been content to focus only on the geometry of the "grid pattern" design in describing Spanish urban planning in the Americas. Others suggest that Spanish planning was strictly derivative from Greek and Roman prescriptions and were concerned with morphology alone (Stanislawski 1947).

Juarros (1823:477-479) gives an excellent description of the manner in which the new capital was planned, the purpose behind the reducciones, and the part that the clergy, as well as the Mexicans and Tlaxcalans, played in providing the agronomic dimension:

...the city is nearly in the centre of the plain, encompassed by 11 suburbs, and these are environed by no less than 31 villages.... The inhabitants of the city derive many advantages from these numerous places; besides the supply of every kind of provisions, they draw plenty of hands for their different works and manufacturies. If a person is in want of bricklayers, he is sure to find then at Icotenango; masons at St. Cristoval the Lower; gardeners at St. Pedro de las Huertas; bakers at Santa Ana.... The inhabitants of Almolonga [founded by the Mexicans and Tlascalan allies of Alvarado] supply the city with fruit of all kinds, either the growth of their own gardens, or procured from other villages towards the mountains, or the sea-shore; Almolongo, and Upper St. Cristoval, furnish all kinds of flowers.... The people of St. Pedro de las Huertas send cauliflowers, cabbages, onions, and every other description of garden vegetables.

The present inhabitants are indebted to the original conquerors for this succession of villages;... for sake of regularity, the whole valley was divided into lots, called Cabellarías and Peonerías; the former 1000 paces long, and 600 broad, the latter, half that quantity.... At this period there were great numbers of the unreclaimed natives wandering about in the forest, and on the mountains, without any kind of subjection or government, who were very detrimental to those who had been already converted. The Spaniards... began to devise means of collecting them together, and establishing them in small villages; this design was still farther promoted by various edicts from the king....

Thus the Indians were "reduced" to satellite villages planned by the Church and Crown. The Mexican and Tlaxcalan immigrants provided an interstitial level of control between the native Guatemalans and the Spaniards. Whether or not the Mexicans and Tlaxcalans served as agronomic specialists and planners is unclear. They did continue to serve as itinerant traders; in part functioning like the Pochteca traders who operated in Guatemala during the Aztec imperium.[18] In this way, the mobility of Guatemalan natives could be regulated and intra-aboriginal trade could be more closely controlled by the Spanish.

LABOR ARRANGEMENTS: 1541-1616

As Jones (1940:13) and others have suggested, the basic Spanish strategies for controlling subjugated peoples originated in Spain, not in the New World. The conquest of the Americas was essentially an extension of the Reconquista of the Iberian Peninsula. Two techniques developed for exacting labor from the subjugated Moors--encomienda and repartimiento--were adapted to the New World situation. Repartimiento involved gang-labor under the supervision of native foremen. Encomiendas provided individual conquistadors with entire villages from which labor could be drawn. The practice of repartimiento became generally subsumed by the encomienda system with time. Therefore, a geographical division of labor by village units became the chief mechanism for exploiting Indian labor. Both systems allowed the Spanish to focus native agriculture in desired directions.

The practices of repartimiento and encomienda were modified under the New Laws. The Crown placed regulations on the sort of labor Indians were permitted to do. The maximization of agro-production was inhibited by certain provisions in the new codes. Through efforts to circumvent these restrictions, field viewers (jueces de milpa) were appointed by the colonists to oversee the Indian's plantings and to dictate the acreage and specific crops to be planted. The practice was outlawed when the Crown learned of this subterfuge, but some forms of the juece de milpa system persisted.

At the end of the sixteenth century, the New Laws continued to discourage chattel enslavement of the Indians. In the absence of out-right slavery, the repartimiento and encomienda systems were intensified. This promoted the continued development of "two Guatemalas." The encomiendas encouraged nucleated and fixed populations directed toward intensive production of food stuffs for the colonists. This legacy is seen in the current tablón systems. In tandem with the further elaboration of Inner Guatemala through encomiendas, the repartimiento system created the basis for the labor migrations to Outer Guatemala. Without channeling the dense populations of Los Altos through repartimientos (later replaced by debt peonage and mandamientos) to the lowlands, cash-cropping of commodity products would have been difficult, if not impossible. While the origins of an Inner/Outer split in Guatemala can be traced to pre-Conquest conditions, the split was intensified under Spanish colonialism.

CRYSTALLIZATION OF THE PATTERN: 1600 TO PRESENT

Through efforts to circumvent the New Laws, the colonists adapted the repartimiento and encomienda systems to the specific demands of production. Under repartimiento, Indians were taken from the highlands to the lowlands to work in the indigo industry. The Crown forbade the use of Indians in direct production of indigo because too many Indians died in this work. African slaves were deemed more suitable. Therefore, Indians under indigo repartimiento were used to grow provisions for the indigo workers.

The colonists' adaptations of the repartimiento and encomienda systems created a measure of autonomy that the Crown eventually found unacceptable. By 1616 new ordinances were enacted to strengthen the central authority. Falling under the general title of "mandamientos," they provided nominal wage payment for Indian labor. It was coercive wage labor because each village had to provide a specific number of individuals to work a given period of time. This was generally sixteen weeks per year. As the system evolved, the governors assumed control of this labor pool and provided gangs of laborers on a commissive basis. At first, the Crown fought the contract labor system. Later, it acquiesced when it became apparent that this could become a new and important source of income for central authority.

National Level

At the national level, the broad patterns of interaction between Inner and Outer Guatemala began to jell during the seventeenth century. Inner Guatemala continued to be a refugium of sorts for the large population of unacculturated Indians. They used intensive agriculture where population levels could sustain it. Inner Guatemala continued to supply the less salubrious Outer Realm with laborers and, in some cases, foodstuffs.

In contradistinction, Ladino efforts at producing commodities for markets beyond the national boundaries were dispersed throughout the lands below the altiplano. Shifting cultivation occurred widely in both realms. In the Outer regions, intensive cultivation was confined to plantation cash-cropping of non-food stuffs. This pattern has strong parallels with the historical development of agriculture in Indonesia as described by Geertz (1963).

Regional Level

At the regional level, the documentation of agricultural conditions and change is spotty. Ecclesiastical records allow some inferences concerning selected locations. There are several descriptions of crops and environmental conditions for the Atitlán Basin. According to the seventeenth century historian, Vazquez (1944:28), the Franciscan order had established its convent in Panajachel by 1600. The first Franciscan establishments were founded in the Valley of Guatemala at the time of the construction of the new capital. He noted that the early administrators of the Panajachel convent were from Almolonga.[19] This may be significant in that Almolonga, as Vazquez and others point out, was settled by Mexicans and Tlaxcalans. The economic specialization of Almolonga combined gardening with fruit raising. One might speculate that the Franciscans at Panajachel replicated the landscape focus and economic specialization that they promoted earlier in Almolonga. Vazquez gave a description of Panajachel in 1689:

The town of San Francisco de Panajachel... is situated on the shores of Lake Atitlán in a cove or bay between two mountain ridges, in a spacious plain whose soil is extremely fertile for all manner of crops, other species, and it has many fruit trees, both from the hot lands and the cold lands. The temperature is moderate but very dry, even though it has an abundance of water from streams as well as the lake which offers easy access. The natives support themselves through commerce with their vegetables and fruits, and the extensive cordage [jarcia] that they make, and also with the bountiful crabs and small fish that crowd the lake.

He also mentioned the extensive inter-village trade that the Panajacheleños carried on by canoe.[20] From this description several things can be suggested. First, two of the crops he

31

mentioned--anise and chian--are not major crops in Panajachel today. Change in crop preferences over time has a long history in
the Basin. Second, the contrast of aridity with the easy access to
water implies that irrigation was used. He also notes that cordage
production and fishing were important activities, which is no
longer the case.

Almost a century later, the situation had not changed per-
ceptibly. Cortes y Larraz (1958:168), writing in 1768-1770, offered this description:

The town produces much fruit and vegetables, but little maize because they lack
the land to grow it. Also they are somewhat short of livestock; though all the
Indians can be well accommodated because the little land there is, is a garden
that provides fruit and vegetables to many pueblos and even to Guatemala
[Antigua--which had its own horticultural villages like Almolonga] and they make
much money from their cords and nets (redes) owing to the fact that they have
abundant fiber which they cultivate well. The town does not have a distinct shape
[not highly nucleated or laid out in a grid-pattern] because the huts are scattered throughout the horticultural plain, and each residence has its own land that
it cultivates.

For the most part, this description fits Panajachel's current condition. Several inferences might be drawn from Cortes y Larraz's
assessment. First, most of the delta land must have been in fruit
or vegetable cultivation, if there was a scarcity of land for corn.
Second, Panajachel's production of horticultural products must
have been considerable if they were sent to Antigua, or regional
production was so specialized that Antigua's satellite villages did
not produce sufficient variety to satisfy its needs.

The population figures that Cortes y Larraz gives are note-
worthy--1,167 Indians and 30 Ladinos. This compares to Vazquez's figure of 800 Indian communicants in 1689. How many
Indians had resisted proselytization is difficult to estimate. Whatever the figure, perhaps as many as 100 or more, the Indian population of Panajachel has been fairly stable since the mid-
seventeenth century.[21] Therefore, it seems likely that Indians
have been practicing intensive agriculture in Panajachel for at
least the past three centuries and perhaps earlier in other parts
of the Basin.

Implications of Pattern

It appears that Panajachel's economy was more diversified in
the seventeenth century than in the early part of this century.
The manufacture of cordage and attendant products from maguey
fiber has not been important in Panajachel during this century.
However, for some time it has been the major activity of San
Pablo; even to the exclusion of agriculture (Rojas Lima 1968:297).

San Pablo was founded in 1727, and as Rojas Lima notes, it is curious that Pableños speak Cakchiquel, though the village resides in the hinterland of San Pedro and Santiago--the two main centers of Tzutujil language use. San Pablo's southern municipal limits delineate the Cakchiquel/Tzutujil language boundary.

This information, together with the date of its founding and the singular nature of its industry, suggests that San Pablo may have been planned by the colonial authorities (civil or religious) in an attempt to further promote a micro-regional division of economic activity within the Basin. Again, this mirrors the scheme the Spanish used to organize settlement and production in the valley of Antigua. In short, San Pablo seems to have preempted Panajachel's production of maguey products either by design or indirectly. In turn, this allowed Panajachel to further expand its horticulture: land once used for maguey could be freed for tablón crops.

Since the seventeenth century the trend seems to be one of increased specialization and intensification of the "geographical divisions of labor" throughout highland Guatemala. This process can be seen in the changes in village crop preferences in the Atitlán area. Until the last few decades, each village cultivated one or more crops as their specialty. Panajachel had pepinos, onions, and garlic; San Antonio anise; San Pedro, garbanzos and maguey; San Marcos, fruit trees; and so on. More recently, as tablones are expanded, the "space efficient" plants like onions and garlic are becoming more prominent, displacing some of the earlier village specialities.

A LADINO ORIGIN OF TABLONES?

The last half of the seventeenth century has been characterized as a time of depression and urban depopulation in Central America (MacLeod 1973:288-329). There were many causes; most were related to the structural crisis in Spanish imperialism (Wallerstein:1974). As a result, Guatemalan urban centers lost population to the countryside. As MacLeod (1973:290) has pointed out: "Many escaped the high prices and food shortages of the cities... by leaving the cities and taking up rural land subsistences or semi-commercial farming." One might suppose that intensive truck gardening using tablones or similar techniques would have been developed by Ladinos during the depression. However, tablones are largely artifacts of Inner Guatemala, and the flight to the countryside was directed toward the vacant lands generally below 1,500 m, especially in the eastern part of the country. Therefore, this possibility does not seem likely.

Other pressures generated by Ladinos may have prompted intensification and specialization of Indian horticulture. Under the repartimiento system, Indians were often made to supply Ladino officials with specific produce. Though nominally illegal, these abuses were wide-spread. This form of extortion was probably similar to the tributary pressure that rural indigenous poor suffered before the Conquest.

Forced migratory labor has been a more enduring source of Ladino pressure on Indians to produce. Under the system of mandamientos that emerged during the 1600s, Indians received nominal monetary compensation for their forced labor.[22] In various forms, this policy has persisted until the last few decades. During the late 1800s, a particularly pernicious adaptation of this practice was created to ensure a seasonal flow of labor from Inner to Outer Guatemala to work on the rapidly expanding coffee plantations. The mandamientos, "or enforced labor not under contract" as Jones (1940:153) remarked, gave way to "forced labor under contract that could be controlled." The mechanism for this "improvement" was a series of statutes outlawing "vagrancy." Indians not owning their own land, or subsisting on the returns from swidden milpas, were particularly vulnerable. Indians engaged in tablón labor, i.e., day-to-day horticultural activities, or permanent activities on permanent sites to which they often had title, were less apt to be charged with vagrancy. Indians with some money could avoid this form of debt peonage. These factors probably helped to sustain Indian interest in the market-oriented production of fruits and vegetables. While the "vagrancy" laws were abolished with the Revolution of 1944, the low return from farming continues to send Indians to the Boca Costa or the lowlands for seasonal work on the plantations. Tablón farmers in Panajachel are generally able to avoid this course. As Hinshaw (1975) has noted, this is because Panajacheleños have a variety of occupations open to them, particularly in the tourist-associated service sector.

Panajachel's location at the core of the tablón district in the Atitlán Basin has allowed it to assume a strong position in the Basin's hierarchy of tablón areas. This advantage is compounded by Panajachel's ability to host a combination of different peasant ecotypes. This seems to give more stability to the tablón system as it functions in Panajachel.[23] As Smith (1972) has implied, there is a core, semi-periphery, and periphery extant within the peasant ecotypes of Guatemala. In her theory, these divisions correspond to the relationship each village or rural area assumes to the "means of consumption" or the articulation of the Ladino market centers. Significantly, Smith's regional construct corresponds with the arrangement Wallerstein (1974) sketched on a global basis: core, semi-periphery, and periphery—all have differing modes of production associated with them.

The question, then, is not whether Ladinos introduced or "invented" tablones, because they almost certainly did not. It is a problem of how much and in what ways have Ladino pressures—

from trade, tribute, and labor extraction--molded tablón culture? The answer seems to be: Quite heavily.

NOTES

[1] There is some documentary evidence that the Mesoameri- cans cultivated varieties of onions and garlic at the time of the Spanish arrival. In his first letter to Carlos V, Cortes said (MacNutt 1908:258): "There is all manner of vegetables that one encounters, especially onions, leeks, garlic, common cress, water cress, forage, sorrel, thistles, and golden thistle." All of these plants are particularly suitable (save the thistles) for raised-bed gardening with permanent flowing water close at hand.

In another connection Juarros (1823:488), citing Fuentes (1690), mentions a cavern at the site of the ancient village of Mexico (a Nahautl placename) and:

The portico, formed of clay, is in some parts entire, and appears to be of the Doric order. Fuentes says, he inquired of some of the old Indians how it has been contrived to give so great a consistency to the clay, and they informed him, it was done by grinding a quantity of onion-seed, and mixing it in the water which the clay was tempered.

This suggest that sometime before the late 1600s Indians were us- ing quantities of onion seeds for building purposes. The Mexican place name and the antiquity of the site raises the question of whether alliaceous crops were being grown in highland Guatemala prior to the Conquest, and, if so, were they introduced from Mexico?

[2] The few lakes in highland Guatemala, which have volcanic origins, are generally deep. The swampy littorals necessary for chinampa cultivation are severely restricted, in the case of Lake Atitlán, and limited in the other lakes.

[3] The flood caused by Hurricane Fifi in November 1974 de- stroyed several cuerdas of tablones, with the principal loss being organic matter. The tablones were in an area only recently con- verted from rubble, and the farmers understood the risks that they were taking. To this extent, tablón expansion is limited in this sector by the threat of flooding, and new plots are developed with this in mind.

[4] For example, the Aztecs placed tobacco in an important po- sition within their religio-pharmacopoeia. As Soustelle (1961:154) remarks:

Tobacco was widely used in medicine and in religious ceremonies. It was thought to have pharmaceutical virtues and a religious value.... The use of the plant among laity in the pre-Cortesian period does not seem to have spread to the common people.

Maya use of tobacco and its importance in pre-Conquest times is well established (Thompson 1964:103-122, Robicsek 1978). Fuerst (1974:192) suggests that the Maya used many of the same psychotropic plants that the central Mexicans used for religious and ceremonial purposes. From the sixteenth century chroniclers Paez Betancor and Pedro de Arboleda (1965) we learn that tobacco was an important crop at the Estancia de San Andrés (associated with Santiago de Atitlán) during the late sixteenth century.

[5] What Soustelle (1961:128) said about the Aztecs fits the Maya attraction of flowers as well: "[They] had a positive passion for flowers; the whole of their lyric poetry is a hymn to flowers, 'which intoxicate' by .their loveliness and their scent." How much the highland Maya cultivated flowers before the Mexican invasions may have bearing on the antiquity of tablones.
Bancroft (1876: vol. 2:349) corroborates Soustelle's view:

Besides regular and extensive plantations of staple products, gardens were common; tastefully laid out and devoted to the cultivation of fruits, vegetables, medicinal herbs, and particularly flowers, of which the Mexicans were very fond, and which were in demand for temple decorations and bouquets.

Brigham (1887:225) felt that the pre-Conquest Maya made use of flowers: "They were very fond of flowers but whether they generally cultivated them, or found them growing spontaneously, we do not know...." Schmid (1969) encountered flowers being grown as the village speciality crop at La Cruz Blanca near San Juan in Sacatepequez. Several varieties are grown and:

The flowers in this region are always grown in tablones or raised beds. Owing to the fairly steep condition of terracing, the tablones being made level with a drop of a foot or more from the level of one to the next below it. The lower sides of the tablón are formed with wet clay. Once the tablones are formed they are not torn down each year, only repaired when necessary. Irrigation is done... (with) water being drawn from the well with a rope and pail, and dumped into a sprinkling can.

La Cruz Blanca is near San Pedro las Huertas; therefore, it may have been one of the garden villages planned by the Spaniards in the sixteenth century.

[6] Wolf (1966:19) defines a peasant ecotype as:

The ecological adaptation of a peasantry consisting of a set of food transfers and a set of devices used to harness inorganic sources of energy to the productive process. Together, these two sets make up a system of energy transfers from the environment to man. Such a system of energy transfers we call an ecotype.

He distinguishes between paleotechnic and neotechnic ecotypes. The first is marked by "the employment of human and animal labor," whereas the second relies on increasing amounts of energy supplied by combustible fuels and "the skills supplied by science."

The concept of an Inner and Outer Guatemala is borrowed from Geertz (1963:14). He uses the distinction to describe the division of Indonesia into two realms. Inner Indonesia exhibits intensive cultivation (primarily sawah or wet rice culture), while Outer Indonesia is the domain of swidden or shifting cultivation and plantations for tropical commodity crop production. A similar division is evident in Guatemala. Highland, or "Inner" Guatemala, is the domain of dense population, intensive cultivation, and production for the Central American, national, and sub-national markets. Lowland, or "Outer" Guatemala, is the realm of shifting cultivation on the one hand and commodity cropping for the world market on the other. Despite important differences between the patterns in Guatemala and in Indonesia, the parallels are striking enough to invite a comparative use of Geertz's model. The highlands above 1,500 m constitute Inner Guatemala (25 percent of the land area); Outer Guatemala encompasses the rest of the country.

[7] The origin of the encomienda system (and indirectly the refinements of the encomienda--repartimiento and mandamientos) can be traced to the feudal reconquest of Spain. In America, however, the encomienda was a capitalist institution as Wallerstein (1974:92) points out:

It was a direct creation of the Crown. Its ideological justification was Christianization. Its chief function was to supply a labor force for the mines and cattle ranches, as well as to raise silk and to supply agricultural products for the encomenderos and the workers in towns and mines.

Jones (1940:13) defines these forms of labor control as follows:

The encomienda divided the natives of named villages among certain conquerors (encomenderos), who where given the right to draw upon the Indians for labor or products.... Repartimiento [was]... the practice of dividing the natives into groups assigned to do tasks under the supervision of native foremen.

The mandamiento was a seventeenth century adaptation of the encomienda system designed to circumvent the protective spirit of the New Laws (1542) (Jones 1940:138). Through the mandamientos, groups of Indians were required to serve a certain number of weeks at work designated by the governors or lesser authorities. In turn, the mandamiento system was reconstituted under the "Vagrancy Laws" in the nineteenth century to provide coerced labor for service on coffee plantations.

[8] Las Casas' population estimates are generally considered excessive. Denevan (1976:291) estimates the population of all of Central America in 1492 at about five million. Regardless of the figures, the disruption was enormous. Bancroft (1876:601) says, "One half of the Cakchiquel population are estimated to have fallen victim to this pestilence...." He was referring to a smallpox epidemic contracted from the Nahua tribes and traders occupying the southwest coast of Guatemala in 1521.

[9] Cakchiquel accounts say that one third to one half of their entire nation was destroyed in two years (1522-1524) (MacLeod 1974:41). The Spaniards saw severely depleted populations; yet they compared them favorably to the dense populations of central Mexico. Pedro de Alvarado's descriptions speak of large populations: "The country [around Quezaltenango]... is as thickly populated as Tlascala and equally cultivated and excessively cold" (Mackie 1924:59). One possible reason why the Spaniards failed to specifically mention intensive agriculture is because the Indians were so extremely enervated. They would have abandoned more ambitious agricultural pursuits, relying instead on feral domesticates, easily gathered auxiliary food sources like ramon nuts (Brosimium alicastrum), and simple swidden cultivation.

[10] Jones (1940:116) remarked:

Satisfactory reports of the amount of the tribute in kind and in labor do not exist.... There is, however, no reason to believe that the Spaniards did not exact what the traffic would bear. The caciques of Atitlán in a petition sent some years after the conquest complained to Philip II that the town paid a tribute of from 400 to 500 slaves to work in the mines, each month, not to count the surrender of 1,400 xiquipiles of cacao, spun and woven cotton, fowls, corn and other articles.

[11] While Wilken (1971:438) is referring to the probable existence of chinampas in the lowlands, his citation of Squire (1858:556) demonstrates the "strategic advantages" of raised-field cultivation: "Spanish horsemen found it impossible to leap over deep ditches surrounding Lacadón lakeside maize fields in the Petén."

[12] Schmid (1969:29) estimated: "Though some garlic is produced in other parts of Guatemala, 80 percent of the garlic is produced in the one region of Aguacatán."

[13] Schmid (1969:29) suggested that the Aguacatán hoe is unique: "The garlic is hoed by men and women with a special narrow hoe that is seen only in this area." Whether or not it is a survival of a pre-Columbian hoe or a post-Conquest adaptation for garlic production is uncertain. The blade is less than one fifth the width of the standard azadón, making it more amenable to construction using pre-Columbian materials (Plate 1).

[14] Juarros (1823:477) mentioned that the inhabitants of Almolonga--one of the satellite towns built around Antigua and settled by Mexicans--supplied the valley with pulque until the governor, Andrés de las Navas, suppressed the practice. Apparently, pulque brewing was a Mexican culture trait brought to Guatemala at the time of the Conquest.

[15] West (1948:43) feels that Iberian irrigation techniques were introduced in parts of Mesoamerica: "Arabic irrigation techniques appear to have been introduced jointly with wheat culture, for both early Spanish and Indian wheat fields in the northern Tarascan area were artificially watered." This raises the question of whether all or part of the tablón techniques involving irrigation were of Arabic origin via Spain. For example, the similarity between traditional Saudi cucumber beds and the modified tablones which are used for raising cucumbers in Panajachel is striking. Nevertheless, parallel and independent invention, rather than diffusion, seems to be the most likely explanation.

[16] Concerning the introduction of new plants, West (1948:46) notes that in the Tarascan area: "The European fruits were brought [in]... early in the 16th century mainly by missionaries. Every mission, church, or convent had its small garden where the priest and Indian neophytes cared for fruit trees and vegetables."

[17] In miniature, this pattern (specialization by village) corresponds to the global spatial ordering of production under what Wallerstein has called "the modern world system." He defines "a world system as one in which there is extensive division of labor... not merely functional--that is, occupational but geographical (1974:349)."

[18] Chapman (1957:120) defines the pochteca as:

Long-distance trader(s) among the Aztec... [who were] various types of full-time professional traders who carried on trading relations exclusively with peoples be-yond the frontier of the Aztec Empire. The only exception to this was the trade with the isolated province of the Empire, the enclave of Xoconusco in the Guatemalan region.

Polanyi (1957) and Chapman see the Pochteca engaged in ritual or redistributive trade (as opposed to monetary market exchange). In the first years of the Conquest the relationship between the Spanish and the Guatemalans was largely tributary. Therefore, it seems reasonable that the Mexicans and the Tlaxcalans (maybe with some Pochteca in their ranks) were agents of the state, not unlike the Pochteca under the Aztecs.

[19] Almolonga, Sacatepéquez, should not be confused with Almolonga, Quezaltenango. The former was the site of the second capital of Guatemala founded by Jorge de Alvarado in 1527. The latter Almolonga is currently a major center of tablón agriculture close to the city of Quezaltenango. Almolonga is a Nahuatl word meaning "where the water flows." The aboriginal name of Almolonga, Quezaltenango, was Bulbuxya, meaning "flowing water" in Cakchiquel. Today Quiché, not Cakchiquel, is the spoken language of Almolonga, Quezaltenango. One might speculate that the valley was colonized by Cakchiquel speakers at some time since the Classic period, perhaps for the purpose of intensive gardening.

[20] In passing, Vazquez mentions that the largest canoes were fourteen varas, or about 13.5 m long. This corresponds to the size of canoes that Lothrop (1929:218) said were once carved from giant cedars from the volcano cloud forest belts. Today the majority of canoes are under 4 m; suggesting a decrease in the importance of large cargo canoes (with the recent construction of roads to some villages, and the use of Ladino motor launches).

[21] The Indian population was 1,585 in 1950 and 2,023 in 1964. The combined population of Indians and Ladinos was 3,268 in 1964. The number of Ladinos has increased disproportionately in recent years. Hinshaw (1975) attributes this to the increased opportunities in the service and building trades.

[22] Wallerstein (1974:91) subsumes this form of labor control and similar schemes under the designation "coerced cash-crop la-bor." He defines it as "a system of agricultural labor control wherein the peasants are required by some legal process enforced by the State to labor at least part of the time on a large domain

producing some product for sale on the world market." Coerced cash-cropping was the predominant mode of production in the semi-periphery (Hispanic America and eastern Europe) of the world economy in the sixteenth century.

[23] Only about 2 percent of the Panajacheleños find it necessary to migrate to the coast to work. In contrast, more than 50 percent of the Catariñecos migrate seasonally to Outer Guatemala for work. More than 85 percent of the Cruzeños (Santa Cruz) must leave their village each year (Hinshaw 1975: 153). These figures correspond to the range of opportunities each village has to support itself through economic specialities. Tablones and tourism permit Panajacheleños to stay in the highlands if they choose.

4
Material Culture

TOOLS

Azadón

The most prominent agricultural implement used by tablón farmers is the <u>azadón</u>, a broad hoe common to the Mesoamerican tropics. Metaphorically, the azadón in the hands of the Meso-american farmer is the analogue of the bow and arrow of the North American Plains Indian. Through a single subsistence arti-fact the contours of a whole cultural complex can be elaborated (Tax 1975).

Assuming that the azadón is essential to the tablón process, the question then arises: Could there have been tablones without the azadón? The tablón-like chinampas of Mexico can be made with the aid of basketry. Similarly, baskets are used as dredges in New Guinea to build raised fields in swampy areas (Serpenti 1965). Presumably this method was used as well in ancient Meso-america. However, tablones are impermanent structures. When made on the sandy and rocky flood plain at Panajachel, some sort of adze or hoe seems imperative.

As Palerm (1967:46) has suggested, the aboriginal adze or coa "comes in many different shapes, usually adapted to soil con-ditions and cultivation systems." If tablones are pre-Columbian, then an implement closely resembling the modern azadón must have been used. Using anything but a facsimile of the azadón would have considerably increased the labor time needed to build tablones. No attempt was made to experiment with alternative techniques.[1] However, based upon actual field observations con-ducted for this study, it is estimated that using any other method would double the time needed. A standard sized tablón takes 8 to 10 hours to construct today. Using less efficient methods, the pre-Columbian farmer could have invested several days at this task. It might be argued that a "semibureaucratized hydro-agricultural" state could have easily commanded the labor power to pursue such projects. The expenditure of energy and time routinely employed in the construction of monumental works of architecture in the pre-Columbian world demonstrates this. But rather than argue this point along the lines set forth by Wittfogel (1957:18), a simpler explanation exists.

Metallurgy was known and used by many pre-Columbian populations for agricultural implements.[2] Bancroft blanketly states that "the coatl, or coa..., was a copper implement with a wooden handle, used somewhat like a hoe is used by modern farmers.... (1876:5, 349)." Donkin (1970) suggests that metal hoes were used throughout much of the Nuclear American highlands. While the shape of these hoes or mattocks were radically different from current adzes and hoes, their function was the same. West (1948:38) has described a survival of this implement among the Tarascans of Mexico. The tarakua has also survived among the mountain people of the Balsas drainage in Guerrero, Mexico.

Donkin (1970) suggests several reasons why the pre-Columbian adze has not been found in Guatemala. First, any metal adze blade existing at the time of Contact would have been melted down by the metal-hungry Spanish and reworked into more familiar implements. Secondly, little archeological work has been done in the Guatemalan highlands, so the remnants of pre-Columbian adzes may well exist, but have simply not been discovered. Finally, he points to the Andean Aymara practice of using animal scapulas for hoe blades. While the highland Maya did not have domesticated Camelids to draw upon, they could have used the bones of a number of wild mammals: deer, manatee (from the coastal lowlands), tapir, and bear. In short, an absence of pre-Columbian hoes in the archeological record should not deny the possibility of pre-Conquest tablón agriculture.

Today, the azadón is considered a necessity for the tablón farmer, who owns one or more. The azadón is available in a variety of sizes and are numerically coded according to the manufacturer. The Collins Company of Connecticut sells a majority of the azadones in Panajachel. Popular sizes are numbers 11.5 (1.6 kg) which sells for Q4.00 and 12.5 (2.1 kg). Both are near the "top of the line" in weight and price. Smaller sizes ($\cong$ .90 kg) are used by young boys. Other brands, like the Corneta from El Salvador, are sold, but preference runs in favor of German brands when they are available; then U.S. makes; and finally the more local brands. Alleged decline in quality with concomitant price increases was a frequent complaint. German azadones were said to last for "years," while the newer brands from North and Central America have to be replaced every few years.

The shafts for azadones come from local trees with Taxisco, oak, and avocado being the preferred stock. They are sold by itinerant merchants plying the paths and streets with carved wood products. The price for a soft wood shaft is about Q0.30, while denser stock sells for about Q0.50 per shaft. The wood for the shafts comes from trees in the monte and represents one of the many uses of the communal forest lands. General access to these lands are apparently open to "outsiders" specializing in the manufacture and sale of wooden handicrafts. Like most economic specializations in native highland Guatemala, these products are also part of the regional and periodic market system. This points out the extreme nature of the craft specialization in this area. Rather than going to the monte himself, the farmer buys a handle from a passing higgler. Moreover, the price amounts to about one half of the standard daily wage for tablón-associated labor. This would seem to emphasize the devotion of the farmer to the marketing systems as it traditionally functions.

Machete

Second to the azadón, the most versatile and ubiquitous tool is the machete. Adams (1957:306) has described the variations of this tool throughout Central America. In Guatemala two basic types are found: the straight form and the hooked variety. The straight-bladed machete is universal in Guatemala, while the curved or hooked form is more restricted. The latter has about the same distribution pattern as the wooden plow. Presumably, hooked machetes are not used in Panajachel. However, minor variations of the straight-bladed machete are considerable. This is a function of aging, and "customizing" machetes through repeated sharpenings. As with all manufactured goods, the farmer makes maximum use of tablón implements, both in function and in longevity. Machetes are often reduced over time to the size of carving knives. When they reach this size they are used in the place of small dibble sticks for weeding and for transplanting vegetables.

Little clearing of secondary vegetation (especially woody plants) is necessary in tablón farming. Therefore, the machete is not as crucial to the tablón process as it is in shifting cultivation. Still, most farmers carry one. Its importance as a weapon should not be underestimated in a country where the legal access to firearms for Indians is severely restricted.

Like azadones, machetes are sold in hardware stores and in the market place. There are various brands. The Collins machete seems to enjoy some degree of preference over the other makes. The price of machetes (and azadones) has nearly doubled in the last few years. Farmers are reluctant to part with a used machete for much less than the price of a new one. Sheaths are used, but the majority of farmers prefer to carry their machetes cradled in their arms to and from their fields. As an adjunct to the machete, wooden brush hooks (garretas) are often used to manipulate brush being cut.

Palito

The small dibble stick, or <u>palito</u>, is used more commonly than the truncated machete in planting tasks. It is used in several specialized tasks, mostly involving the transplanting of onions. They are generally about 85 cm long with the wood usually being the same types seen in azadón handles. One end of the palito is sharpened, and it is wielded at acute angles from the body to the ground. Most often these dibble sticks are used for weeding and for removing seedlings and starts from seed beds before transplanting. The palito is also used to perforate the tablón top for injection of chemical fertilizers below the soil surface. Additional uses include punching holes for planting sweet potato cuttings in tablón sides and for aerating tablón tops for better drainage.

Other digging and cutting implements are employed, such as the mattock and light-weight garden hoe, but only infrequently. These may be examples of tablón farmers adopting the use of European garden tools from their occasional stints as flower gardeners and lawnsmen for the vacation-home owners. In general, the azadón serves a variety of tasks, obviating the need for many tools common in Euro-American gardening.

Dray Implements

A variety of carrying implements and containers are used by tablón cultivators. They also serve as the basis for volumetric measurement. The <u>palangana</u>, or small tin wash basin, is a versatile implement which varies in size from 15 to 40 cm in diameter. The primary use for the palangana is in splash irrigation. This is done with a medium sized palangana (26 cm). Apparently, this particular palangana is the only implement used in splash irrigating today. At one time, bottle gourds (<u>Lagenaria</u> spp.) were used for irrigation. Although the hollow bottle gourds are sold in the markets, their function as an irrigation implement has been replaced by their metal equivalents. Plastic replicas of the metal palanganas are available, but they do not seem to be replacing the metal pans.

Palanganas are also used for carrying activities. Seeds, starts, peels, bulbs, and cuttings are carried in palanganas before they are set or sown. The starts used in transplanting onions, beets, cabbage, etc., are carried from the seed beds to the tablones in the basins. Finally, women sometimes wear palanganas on their heads in the fields if they lack their traditional cloth head pieces. Medium-sized palanganas sell for about Q0.35, and they are generally made in the Orient. Enough discarded palanganas are found around tablones to suggest that these pans must be replaced frequently.

Woven baskets are used in harvesting many vegetables. These baskets may approach 1 m in width. They are used in collecting sweet potatoes, cintul roots, and other rhizopodous crops. Onions harvested for market are collected in baskets before they are transferred to woven nets for shipment. As mentioned above, baskets are not used in constructing tablones, nor are they used for cleaning the irrigation ditches. This is in contradistinction to the practice of raised-field farming in poorly drained terrain in parts of the tropics, where basketry is widely used for construction and maintenance. Like other handicrafts of the highlands, woven baskets are the product of economic specialization and reach the tablones through regional marketing networks.

Redes, or woven nets, compliment the use of basketry in the production and processing of tablón crops. These nets are made exclusively in the neighboring lake town of San Pablo west of Panajachel. Redes are woven from maguey cactus fiber produced in San Pablo; the size varing according to the length and number of pitas, or cords, used in their construction. Three sizes were mentioned in conversations with Pableños: 45, 50, and 55 "pita," or "maya," a denomination that simply indicates how many cords the net contains. When filled to reasonable capacity with avocados or onions, the red assumes the shape of a large Ottoman foot rest. A full bundle of avocados weighing 125 kg has circumferences of about 285 by 216 cm. Redes sell for Q0.80 per pair; a pair is termed a carga. Redes are used for a variety of tablón-associated tasks, but all are connected with conveying produce (mostly onions, cabbage, beets, corn) to market. Like other Cakchiquel dray implements, the red is often used as a measurement of quantity. For example, middle men buy redes of avocados for a set price, and later sell the avocados (or cabbage) by the red as a unit. Several functions are served with one implement.

Besides the red, other implements are used for carrying produce or materials. Intermediate in size between baskets and redes are large "tin" containers of several gallons capacity. These latas are used for transporting animal manures and top soil. They also serve as standards of volumetric measurement. As an example, chicken manure is commonly sold by the lata.

Slightly larger than the latas are the wooden boxes of varying sizes made by Indian craftsmen for all manner of conveyance in the highlands. These trunks are used to carry personal effects in the case of migration (seasonal or permanent). They are also used in tablón farming. One farmer from a hill community traveled each day to the flood plain to work his tablones. He carried a rather large box on his back using a tumpline, and scavenged top soil along the way. He considered the 0.5 square meters of top soil collected periodically as an important way to appreciate his "soil capital."

Ordinary flour sacks are sometimes used in place of boxes or redes. They serve several functions. Organic compost is often sold to tablón cultivators by the sack. In addition, the occasional shipments of sheep manure that come to the Lake Basin from the tierra fría are transported with cloth sacks. Pine needles are also transported by cloth sacks from higher elevations to Panajachel. They are used as a covering on the seed beds devoted to onion and beet nurture.

The final major implement used for conveyance is the wheelbarrow. Because craftsmen make a local variety of wheelbarrow, little or no market exists for manufactured wheelbarrows. These local constructions are made entirely of wood, save for a few bolts or screws. They are essentially two handles attached to an axle with a wooden wheel. The handles act as supports for a flat wooden deck or platform, rather than the more usual bucket or box of Euro-American design. The use of the local wheelbarrow seems to be restricted to Indians expressing some degree of Ladinoization. Indians in pure dress use a box and tumpline for hauling rocks from their fields, produce to market, or firewood from the river. Tentatively, it might be suggested that the use of wheelbarrows, as opposed to tumplines, offers yet another indicational trait for ranking Indians according to their degree of cultural modification.

If the wheelbarrow is a rarity among the "purer" Indians of Panajachel, then the tumpline is a versatile and universal substitute. The tumpline is a leather strap hitched around a package, or red, of cargo and worn across the forehead. It has been suggested that the use of the tumpline (which requires a deep bend forward at the waist for the wearer) has even influenced the way in which the highland Indian moves. The Indian demonstrates a characteristic gait with short, quick steps taken at a half walk, half run. This style of locomotion seems well suited for heavy loads requiring stooped posture, and is seen throughout the region. It also stands to reason that the wheelbarrow is not a serviceable piece of equipment beyond the level confines of the village. The tumpline is predominant in beyond-the-village transport.

Other Artifacts

A number of other implements are used by various farmers in work related to tablón culture, but they seldom appear with the frequency of the equipment mentioned above. Some farmers spray their tablones with insecticides and herbicides. They use plunger-tank spraying equipment, or even aerosal spray cans manufactured by international petroleum corporations for the Latin American trade. How widespread the use of chemical spraying has become in Panajachel is difficult to estimate. At least one store in the departmental capital of Sololá specializes in these chemicals for the Indian market. The current world monetary and inflationary crisis may slow the acceptance of these agents by Indian farmers. If the example of chemical fertilizer is comparable, then many Indians will be forced to abandon practices that they only recently adopted.

DOMESTICATED ANIMALS

Role of Domestic Animals

Tax (1953:117) summed up the role of domestic animals in Panajachel by saying, "Compared to agriculture, animal husbandry is extremely unimportant." Taking his analysis a step farther, he said that animal husbandry "is uneconomical." One of the strengths of Tax's study is the lengths to which he (and his indefatigable field assistant, Juan de Rosales) went in collecting data on the material culture of the town. His study of domestic animals is no exception. However, to say that animal husbandry is categorically "uneconomical" in the context of Panajachel reduces the people and their animals to taxon economicus. Recent work in ecologically inspired anthropology has suggested that what might seem "uneconomical" or "irrational" as regards domestic animals and agroecosystems often turns out to be quite rational when viewed within the total system (Harris 1974; Rappaport 1968).

Tax reported that in 1940 no family depended on raising animals for a living. Moreover, only the "wealthiest" Indian families could afford to keep larger animals such as cattle, sheep, horse, and mules. In addition he saw a direct correlation between the amount of land owned and the number and the kinds of animals kept. Tax even extended this generalization to include small animal raising. Despite the local adage "that every housewife has her chicken," Tax found that only 60 percent of the families kept chickens. Tax concluded that maintaining fowl was dyseconomical. But he admitted that chickens, as well as other animals, represent a form of saving that could be converted to cash at the stroke of an axe. Moreover, chickens represent a source of protein that can be tapped in case of illness, which is often done. To conclude that animal husbandry is dyseconomical is a short-sighted assessment.

Seen in broader terms, the maintenance of fowl, or small animals like shoats, appears quite rational for the following reasons. First, there is the question of what might be termed "socio-homeostasis." The fact that the wealthy upper quarter of the Indian community maintains most of the larger animals is not surprising. However, if animal husbandry is generally uneconomical in the short run, then this may operate as a leveling device for regulating wealth at the community level. Wolf

(1955) has argued this point in general terms regarding the functional basis of hierarchy maintenance among Maya peoples.

At a more expansive level, animal husbandry must be seen diachronically. It is to the advantage of the poorer peoples to have some source of quickly convertible wealth on hand to avoid situations that otherwise might throw them into debt peonage. The roots of this motivation for animal keeping can be traced back to the decades immediately after the Conquest. Finally, ecological reasons for keeping large animals are obvious, if we keep in mind that Panajachel is preeminently a town devoted to intensive horti- culture. Large animals provide much needed manure for soil build- ing. Seen in this light, the farming system as a whole benefits from the presence of animals, even though some of the individual owners may expend more money in maintaining them than they receive in cash upon selling them.

Fowl

While chickens are the most numerous domestic animals kept by Panajacheleños, they are found in only about one half of the households. Their importance as a protein source seems to be less than in rural Mexico. This probably reflects the fact that Indians throughout Mesoamerica are less involved with domestic animals than their Ladino counterparts. The debate over the date of the introduction of the chicken into the New World is quite open (Carter 1971). Presumably the relative disinterest in chicken hus- bandry among the more traditional Indians may be a culture trait lending support to arguments favoring post-Conquest introduc- tion. In any event, chickens became an obligatory part of the Indians' agricultural landscape soon after the Conquest. This is attributable to the Spanish tributary pressure that forced Indians to raise fowl.[3] Here we see that the adage Tax felt to be so fanciful actually has historical justification. The fact that today every "ama de casa" does not have her chickens may be an eco- nomically sound practice as Tax insists, or simply the result of the relaxation of tributary pressures in modern times.

Today, chickens are raised for direct consumption, for marketing, for ceremonial purposes, and for agricultural applications. They are generally allowed to forage, feeding on midden material or insect life found in the leaf litter that accumulates in most "huertas de casa." These "forested" house gardens simulate an environment closely akin to one in which chickens originally evolved. From an ecological perspective, chickens kept in these surroundings represent a prudent use of resources. If chickens were fed too much maize or other grains popular for human consumption, then there might be a question about their economic utility. However, it appears that chickens raised in house compounds with a cuerda or two of trees are a judicious use of resources.

Feed that does not come directly from the chicken's own foraging comes from weedy plants grown in tablones for this and other ends. Apazote (Chenopodum edulis) is one of the common sources of feed for poultry. In addition, chickens have access to a recently expanding food source. "Organic" garbage discarded by the affluent vacationers is tapped and probably represents a substantial increase over the amount of refuse that was committed to middens by the Indian population of Panajachel in the past. A final fact may have some bearing on the persistence of chicken raising. Today eggs sell for six times more than they did in 1936. In comparison, the price of onions has risen only threefold during the same period. While this rudimentary use of price equivalencies may do violence to the concept, it should point out that keeping chickens has become, on the surface, more renumerative, whatever the long-term case may be.

Tax reported an "absence" of turkeys in his study (1953: 118). However, at the time of the current study, turkeys were found in a number of households. They were kept by Indians with strong local roots, as well as by Indians of "foreign" descent. Therefore, they do not seem to be the result of recent diffusion from the outside. Like chickens, turkeys are allowed to forage within the house compounds, but they often range into neighboring tablones where they are an occasional nuisance. In these cases turkeys are a source of amusement. Time is taken out of tablón-related tasks to throw pebbles at the turkeys and drive them back to their owner's huerta. Also like chickens, turkeys are fed on weedy forage, and do not represent a significant burden on the owner in terms of feed costs. Their care and maintenance take a negligible amount of time with most of the work being done by children.

The importance of turkeys, even more than chickens, would seem to lie in their role as reserve sources of money and protein in times of need. In addition, they exploit niches within an environment of closely spaced residential and horticultural sites that precludes the possibility of allowing larger animals freedom of movement. This point may have pre-Columbian implications as well, for most Mesoamericans made little or no effort to domesticate the various species of Cervids and Ovids at hand, while they may have domesticated the turkey. The turkey is an integral part of the maize, amaranth, bean, chile and squash complex of cultigens; deer and sheep might have been a superfluous or even a disequilibrating addition to the complex after it evolved. In light of the antiquity of the turkey as a Mesoamerican domesticate, the absence of turkeys from Panajachel during the 1930s is curious. Perhaps they disappeared in that decade of hard times, reappearing in the generally more prosperous times that have followed.

Somewhat less in evidence than Galliformes are domestic ducks and pigeons. The pigeons include domestic races, as well as wild species of pigeons and doves, that are kept as pets and for squabs. These latter palomas and palomitas del monte are kept in semi-domesticated captivity. This raises the possibility that semi-domesticated Columbiformes provided the aboriginal Maya with a food source that has not been pointed out in the literature. Sauer (1966:58) mentioned the vast wild migratory avian resources available to the pre-hispanic populations of Hispanola. Among these Caribbean food sources were the white-crowned pigeon (Columba luecocephala L.). They flocked in the millions, if not billions, in earlier times, much like the extinct passenger pigeon did in North America. Perhaps the aboriginal highland Maya were able to tap large populations of red-billed pigeons (Columba flavirostris) or other wild pigeons. Red-billed pigeons are still fairly common in the forests around Panajachel today.

The ducks seen in Panajachel are moscovy and other common domesticated stock. Very small ponds are made by damming the main irrigation ditches or digging small pools and feeding these receptacles with water from the cunetas. Ducks play only a small role in the total poultry-raising activities of the village. This might seem surprising at first glance, given the access Panajachel enjoys to the lake. However, the lake drops off too precipitously to be of much utility in supporting domesticated ducks.

Small Mammals

Lagoculture is similar in function if not incidence to poultry raising in Panajachel.[4] Tax took only passing mention to rabbits and guinea pigs in the village during the 1930s, but today several Indian families raise them. One Indian who avidly raises rabbits is also a guardian for one of the larger villas on the Jucanyá side of the river. His occupation as watchman includes lawn work. This raised the question of whether lagoculture's appearance in Panajachel is partially a result of the increase of leisure homes in the town. Lawn clippings provide a significant portion of the feed given to the town's domestic rabbits. This may represent still another ecological adaptation with direct subsistence import for the Indian population, related to the changing residential patterns of Panajachel.[5]

Pigs

Hogs are generally in evidence in Panajachel, and as Tax (1953:22) remarked, swine culture is the most developed aspect of large animal husbandry among most highland Indians. This may have a direct historical explanation. From the earliest post-Conquest times, <u>ganado</u> <u>mayor</u> (horses, mules, cattle and oxen) were regulated by the Spanish authorities, and Indians were not allowed to own them (Gibson 1952:156). Therefore, Indians adopted the unrestricted <u>ganado</u> <u>minor</u> (hogs, sheep, goats) and specialized in their production. Despite the popularity of swine raising among Indians, they do not concern themselves with all aspects of the practice. Ladinos do the gelding and butchering. Indians buy shoats from itinerant swine herds that travel down from Chichicastenango and other tierra fria towns. The Panajacheleños keep the small pigs for maturation and sale.

Even though pig raising is popular with the Panajacheleños, pigs do not assume a central subsistence role. In some more primitive horticultural societies, like those of the southwest Pacific (Rappaport 1968) or parts of Central America (Bergman 1969), pigs are quite important. However, pigs under 20 kg do serve some function directly in the tablón system. These smaller pigs are allowed to range freely in and among the tablones. They seem to ignore the produce bound for market and feed on weeds. How they are restrained from rooting in the tablones or eating vegetables and root crops remains a mystery to this observer. Pigs larger than 20 kg could not be allowed access to the tablones for fear of damaging the tablón walls and "ears," if for no other reason. Larger pigs are kept close at hand, usually within pens in the house compounds.

In this study, no attempt was made to examine the profitability of pig raising. Tax (1953:118) made a convincing case for the fact that more corn is normally spent on raising a hog than is realized by the hog's eventual sale. In this case, the argument put forth by Tax seems reasonable given that land is considerably circumscribed in Panajachel. In a tropical forest situation involving extensive or moderately intensive farming, swine culture represents a rational use of resources (Brookfield 1970; Clarke 1971; Rappaport 1967; Bergman 1969). On the flood plain of Panajachel, with its carefully tended and sculpted tablones, this could hardly be the case. Perhaps then, the explanation for less than full participation in swine keeping by Panajacheleños rests more with the "dysecological," than the "dyseconomic," drawbacks for the villagers. Pigs do provide some manure for the tablones, but how much or what percentage of the total animal wastes applied as fertilizer come from hogs remains outside of the scope of this study.

Goats and Sheep

Goats and sheep are raised by some Indians in Panajachel, but do not represent a significant investment of time and energy. This contrasts with the situation in many of the villages of the tierra fria. In Panajachel, goats and sheep are allowed to graze and browse along the river banks in areas that are sometimes flooded and often forested. Therefore, they draw their sustenance from areas of marginal utility. The locally generated sheep and goat dung is not consciously applied to the tablones. Because these animals are few in number, systematic efforts at collecting their droppings is considered a waste of time. Sheep and goats were never seen close to tablones and in other ways seem to be tangential to the system.[6]

Cattle

Cattle were found in dispersed locations throughout Panajachel. At least six, perhaps more, families had milk cows. This is about the same number Tax reported (1953:119). Cows, usually kept in corrals of barbed wire or birch logs, are taken out to pasture in the various grassy spots within the graveled flood plain. Other grazing spots include the right-of-way edges of the ten or so east-west oriented paths that cross the river. These areas are nominally communal land, but some respect for "territorial propinquity" concerning use patterns is observed. Also, only the more wealthy Indians can afford to keep cattle. Thus some degree of regulation of their numbers in relation to carrying capacity may be achieved in this way.

Farmers owning cattle--three or four head is the maximum-- are supplied with a fertilizer source well respected by the tablón farmers. Some farmers spoke of the advantages of having ganado (cattle) in this current era of inflationary fertilizer prices. At least one farmer, who owns cattle and uses the manure on his tab- lones in the face of peer pressure to switch to chemical ferti- lizers, now feels vindicated. Similarly, cattle owners are in a position to command a good price for their manure should they decide to expand their cattle holdings. However, this seems unlikely given the limited space for pasturage. The number of cattle that the flood plain can support may have reached its limit some time ago and is presently in a "steady-state" condition. Most of the milk products that are marketed in Panajachel come from outside the village. Much of the raw-milk cheese that is peddled in the street comes from the Cabrera finca on the steep land directly north of the town. The town's milk comes from commercial dairy farms in the highlands east of Panajachel. Committing more land to cattle for the sake of greater manure returns would probably benefit the tablón industry, but this is not likely to occur without a major change in land use patterns.

Apiculture

Bee keeping, a major Mesoamerican cultural trait, has attracted the interest of geographers (Sapper 1935-36:183-198; Nordenskiold 1919; Bennett 1964). The ancient Maya made their ceremonial drink, balche, from fermented honey collected from domesticated stingless bees (Gann and Thompson 1931:193). Honey was an important item of trade as well.

Today, bee keeping in Panajachel is declining or has collapsed. Formerly, apiculture was a popular way to supplement income. For reasons that the Panajacheleños were at a loss to explain, the bees have disappeared. Indirectly, one possible explanation has been suggested by West (1948:51). He reported that in the Tarascan area of Mexico, bees are not able to withstand the environmental disruptions caused by volcanic activity. Another alternative is the possibility that increased popularity of chemical pesticides and insecticides is exacting a toll on the bee populations around Panajachel. Bee keeping may represent one of the earliest successes in animal domestication in Mesoamerica (Bennett 1964). The collapse of this activity in areas enjoying considerable continuity over time may offer an "early warning" indicator of environmental stress and degradation. While no firm evidence exists regarding the amount of biocides currently circulating in Panajachel's environs, the disappearance of the town's bees may signal unacceptable levels.

Companion Animals

A final aspect of animal husbandry and its intersection with Panajachel's tablón culture concerns companion animals--chiefly dogs and cats. As Tax remarked (1953:121), dogs and cats represent little commercial value, but they are kept in considerable numbers as pets. In general, dogs remain close to their respective house compounds. They have little bearing on the farming practices of the area, save as consumers of some of the produce. Cats, on the other hand, are aids in policing the fields and discouraging pests. Cats hunt among the tablones for small animals. The opposite is true for dogs; they are considered "pests." Sharpened stakes are sometimes placed around the perimeter of tablones to ward off roaming dogs.

CONSTRUCTIONS

Outbuildings

A variety of ranchitos (agriculturally-oriented outbuildings) are constructed within the tablón plats. They range from small replicas of residential houses (Plate 2), to lean-tos of thatch and banana leaves. A few semi-residential ranchitos, owned and maintained mostly by the Jorgeños, are made of adobe with tile or metal roofs. The farmers occasionally stay in them overnight when they do not want to make the trip back to San Jorge. These well-built cabins also serve as tool sheds and cook shacks. The element of security that better-built sheds provide should not be overlooked. Social relations tend to be rather atomistic among tablón farmers. Jorgeños feel more secure leaving their possessions in locked ranchitos when away from their plots. Ranchitos can be built quite cheaply. One Jorgeño built a ranchito for Q32.30. The bricks were made from mud at hand. The major expense was posts, beam, and straw for the roof. If he had paid a bricklayer to do the job with purchased bricks, he estimated the cost would have been Q185.00.

Less elaborate structures are built by Panajacheleños for the same purposes, save, of course, for overnight stays. Rather than going to the expense or trouble of building with adobe, the Panajacheleño usually uses cane poles or boards. Sheet metal roofs are common on these structures because they are seldom larger than 1.7 by 2.5 m. These outbuildings are not used for storing produce, though some articles of value like palanganas and light tools are left overnight in these sheds. The main function seems to be for rain protection during cloud bursts and for organizing the mid-day meal. Sometimes the ranchito serves as a place to hold a costumbre when a zahorín is retained for crop rituals.

Fencing and Walls

Fencing in individual tablones, or even fencing for property line demarcation, is not common in the open tablón areas. On the other hand, fencing and walls in residential areas (which often include a few tablones near the house compound) is common, if not universal. Ramparts of rock and dirt and concrete walls are used in various parts of the delta for flood protection.

The fencing used for tablones, though uncommon, almost always consists of cane poles lashed together in rectangular grids. The function of these fences is to keep traffic (which normally follows the irrigation ditches in the open tablón areas) off a corner or end of a tablón. Seed beds, on the other hand, almost always have cane fencing around them (Plate 6). Birch log fencing or barbed wire is used in the upper west side of the delta where cattle are most numerous. Some rock walls are used in conjunction with tablones, but these are usually repositories for stones removed from tablones. Rock and mortar walls exist, but they are rare.

Hedging is a well developed practice in the residential areas of Panajachel. A variety of plants are used as living walls. Some of them have striking flowers, adding to the town's color. While hedges have only marginal utility in regard to the tablón system itself (by harboring small animals and birds that prey on harmful insects), the two major concrete walls of the town are essential to the system as it currently exists.

The walls were built on each side of the river in the early 1950s, after the major flood of 1949. In most places the walls are 2-3 m high, except where covered by alluvial debris in select spots. The walls are 1.4 m thick and provide a convenient north/ south access way on both sides of the river. Daily traffic on the walls is moderately heavy and traveled mostly by Indians. Ladinos prefer the regular paths or roads. New tablón land was developed as a result of the security that the walls provided. In several places, however, the walls have been partially undermined by recent flooding. The farmers at these sections are convinced that future flooding will do major damage. They seem resigned to this eventuality, and made no mention of attempting repairs on their own.

NOTES

[1] As Puleston (1971:322) remarked: "Experimentation in archeological research is a rarely used source of inspiration and data of much greater potential than is generally realized." Puleston (1976) experimented with irrigated raised-field systems in Belize in attempts to simulate ancient Maya subsistence patterns.

[2] García Pelaez (1968:47) stated: "In working their farms, the ancient Guatemalans used the axe and the hoe. The first to cut the forests, the second to till the soil." Bishop Las Cacas, discussing Cuscatlán (now San Salvador) in the Destrucción... (Chapter 8), mentions that the Indians presented Alvarado with a very large quantity of copper axes. Remesal (1932, book XI, chapter 19) says that the Indians of Vera Paz also used copper.

[3] As Brand (1951:152) suggested, "At one time the requirement was that every tributary Indian should raise at least 12 chickens and 1 turkey gobbler and 6 turkey hens."

[4] I suggest the term "lagoculture" to describe the activities connected with rabbit, guinea pig or other small mammal raising for subsistence purposes.

[5] The role of lagoculture in the pre-Columbian food complex of Mesoamerica has received spotty attention. García Granados

(1937:111) suggests that small rabbits (probably semi-domesticated) were kept by the Aztecs for eating.

Apparently in the New World, only the Andean peoples developed lagoculture to the status of a major activity (Gade 1967). Gall (1968:65) raised the possibility that the famous mute dogs of the Mexicans and Maya were actually domesticated pacas (Cuniculus paca nelsoni). In any event Sauer (1959:221) emphasized that, domesticated or not, hyristocomorph rodents provided an excellent and abundant source of protein for tropical Amerinds, particularly those planters relying heavily on root and tuber crops.

[6] As MacLeod (1973:124) and others have pointed out, the expansion of stock raising, especially sheep and goats, marked the decline of agriculture in many parts of Spain from the Middle Ages on. This same process was recapitulated in Central America during the sixteenth and seventeenth centuries.

5
Crops

We landed, after a water journey of nearly three hours, at Tzanjujú, where Indians, clad in black clothes and queer black hats, solemnly raked at beautifully-
grown rows of onions and beets with rills of water rippling between.

Lillian E. Elliot, 1925

Central America

CROP SUCCESSIONS AND ASSOCIATIONS

In connection with this present study, a majority of the several thousand tablones in Panajachel were surveyed for crop character, with some 1,300 being surveyed at the beginning of the
dry season (late November and early December). Pains were taken
to record all plant species of direct social or economic value in
each of the tablones within the transect. As the survey was con-
ducted, it became apparent that there was more crop diversity
within tablones than has been previously thought. Tax (1953),
McBryde (1947) and most recently Veblen (1974) underestimated
the special variety (number of species) to be found within tablón
complexes.

Veblen argues that Wilken (1971) is in error in suggesting
that tablones may be of pre-Columbian origin because of Wilken's
statement that "many of the plants so tended are native." Veblen
apparently assumed that because the most conspicuous vegetables
bound for market were Old World introductions, there must have
been a paucity of native plants. While the crops grown at Almolonga and Zunil (the basis of Veblen's assertion) were not surveyed by this author, surveys from both Panajachel and from
Sololá (which more closely approximates the ecological conditions
of Almolonga and Zunil than Panajachel) suggested that Old World
crops do predominate, though not to the extent implied by Veblen
(1974:313).

Panajachel enjoys favorable growing conditions for a wide variety of crop plants. As McBryde suggested, Panajachel is close
to the absolute altitudinal and climatic limit for many of the trop-

ical and subtropical fruits and vegetables that grow very well there. Evidence suggests that many economically useful plants produce best in conditions near the limits of their various ecological tolerances (McBryde 1947:147).

Citrus fruits illustrate this principal. At this latitude, the Atitlán Basin provides a sheltered reserve above the normal altitudinal limits for citrus growing (1,000 m). Three advantages are apparent. First, the climate at 1,600 m more closely approximates the extratropical habitat of citrus trees. Second, "all crops grown near the limit of their 'cold margin' may have slower maturation," and thereby enhance qualities found to be desirable by humans-- i.e., taste, texture, etc. (McBryde 1947:147). Third, disease and pest problems are minimized at the higher (colder) margins of tolerance. Therefore the number of native and non-native crops found in Panajachel is probably greater than in the tablón centers at higher elevations--e.g., Almolonga and Zunil. Panajachel not only accommodates many plants at the upper limits of their ranges, but also hosts several crops growing at the lower limits of their ranges. The white potato is an example of a highland crop found at the lower limits of its range. In short, Panajachel is propitiously located at the juncture of several climatic zones and offers viable habitat for a maximum of different crops. As a result, any argument that denies tablones a pre-Columbian origin on the basis of crop types alone is called into question.

During the current study, farmers were interviewed on various occasions. They were asked to point out what crops were growing in the tablones directly before them. Often they would fail to mention native crop plants used for pot herbs, condiments, and forage for their fowl. When asked about these plants they acted as if these indigenous (often weedy) plants were a "given." The farmers were largely concerned with pointing out crops destined for market, and not plants used for direct consumption.

This separation between crops for market and plants for home use takes on another distinction. On several occasions Indians made forceful and somewhat bitter remarks about Old World market crops being "Ladino" items. They felt that they were being forced to grow these crops, especially onions, for a market over whose prices they had no control, and for returns that often amounted to gratuitous production with no cash returns. The alternative to growing onions would be wage labor on coastal plantations, something they consider even less desirable. In contrast to the way they felt about "Ladino" crops, these farmers demonstrated an affinity for a variety of native plants, such as semi-cultivated greens, sweet potatoes, and manioc. Moreover, they said that the cultivation of "native" plants around the margins of their tablones allowed them to subsist. Perhaps the strong feelings of a few articulate and vocal Indians do not constitute a general attitude among tablón cultivators at large. However, it does exist in the minds of some, and, as such, deserves the attention of an ethnoscientific study directed at the question of native attitudes toward native and introduced crops. For our purposes, the fact that the market receives some, but not all, tablón crops should be acknowledged in attempts that seek to generalize about the degree of "nativeness" of tablón crop complexes.

In general, the cultural separation that divides the whole of Guatemala into two camps--Indian and non-Indian--is evident in the way the Indians classify their tablón crops. The cultural/ historical bifurcation that is evident in all aspects of Guatemalan national life is also evident in cultivated landscapes. As previously mentioned, from the time of the Spanish Conquest, if not before, the country has been divided into two distinct realms. The intensification of Indian agriculture in highland Guatemala is a reflection of this split. Earlier in this century geographers were concerned with describing the "personalities" of regions and countries (Dunbar 1974). Sauer (1941), who rendered a portrait of Mexico in this tradition, chose to frame the essay around the historical division between northern and southern Mexico. Should a similar essay be written on Guatemala, it is clear that the focus would necessarily examine the split between Inner and Outer Guatemala. It is an attitudinal, altitudinal, economic, and historical division that is as clearly marked as any national division within the Hemisphere. Not surprisingly, it is expressed in the crops grown by Indian farmers on the flood plain of the Río Panajachel.

Despite the rather random way in which plots are committed to particular plants, there is recognition that intercropping has advantages over monocropping. While no farmer gave specific reasons for the various associational patterns of intercropping, there is a preference for some intercropping. About 27 percent of the tablones had moderate intercropping (four or more crops), while 6 percent had eight or more different crops. The economic advantages of planting starch staples for personal consumption is generally recognized. The practice also applies to a variety of semi-domesticates that serve as greens, condiments, and medicinals. When pressed on whether some plants are intercropped for their value as insect repellents, farmers either misunderstood the question or answered negatively.

Information on successions was difficult to obtain because it demands the monitoring of individual tablones for periods of a year or more. Seasonal factors exercise restraints on which crops are grown and in what order of succession. Regarding the choice of one crop over another--for example, beans or garlic, when either would be appropriate according to their seasonal limitations--the farmers were particularly vague. Apparently, they do not decide which vegetable will be the predominant crop in each tablón until practically the moment of planting. It is hard to say in this case what motivation is involved. Non-Indians often receive noncommital answers, or even blatantly wrong answers, to questions the farmers consider foolish. Perhaps the farmers are reluctant to give out information on planting for fear that it could be used in some way to manipulate the market to their detriment at a later date. This fear was indirectly stated several times. More likely, the cultivator does not have a conscious yearly plan for planting. The succession of crops within the tablón complex simply follows a pattern according to season, surplus seeds left over from previous plantings, and especially the price of different seeds at the moment of purchase for immediate or future planting.

Seasonal limitations on planting certain crops are observed, but not as rigidly as Tax imagined (1953:47). To be sure, corn is planted in late May at the beginning of the wet season. Garlic generally succeeds milpa on many tablones at the beginning of the dry season in October. However, suggesting that all of the "miscellaneous" vegetables are grown from November to June does not describe conditions as they exist today. Similarly, Tax found bush beans, as well as vine beans, growing only from November to June. While there is a marked increase in ground bean planting after the rainy season (May/October), both kinds of beans can be found growing throughout the year in Panajachel. With enough labor and water, any crop can be grown during the dry season. The decision not to grow milpa in the dry months stems largely from the disproportionate amount of work it would take to irrigate the corn and the decreased number of tablones that could be used for cash crops as well as subsistence greens. These "peripheral" greens grow better in the open micro-habitat of the tablón tops; intercropping them with maize is less effective.

CROP TYPES

As suggested before, three generalizations are implicit in the literature concerning the crops grown on tablones (McBryde 1947; Tax 1953; Veblen 1974). First, the "monocultural" character of Panajachel's tablón industry is stressed. Accordingly, onions and garlic are considered "primate crops." This was empirically confirmed by studies done for this study, but this view simplifies the complexity of the total system. As shown below, considerable diversity exists within the system.

The second implication suggests that tablón crops are primarily European introductions. This notion can also be disputed, as we have previously shown. A related supposition is: Given that most of the crops are European in origin, it is unlikely that tablones are pre-Columbian artifacts. This can also be disputed on the basis of current crop composition.

The third generalization arises more from lack of clarity than from erroneously held assumptions. Some observers have assumed that maize is separated spatially from the tablón complex. Maize is widely grown on tablones throughout the delta during the wet season. In essence, a maize/vegetable seasonal rotation is practiced in the delta. Therefore, the strict maize/vegetable/coffee trichotomy of delta land use is inaccurate. Only coffee is distinctively separated from the tablón complex, and this separation is not absolute. Tablones are co-joined in places and in certain phases with coffee production.

A list of common crops is given in Table A.1. Beyond information in the table on incidence, uses, and so forth, certain crops appear in associational patterns and successional patterns. From this information we can get a preliminary idea of the overall nature of tablón production.

THE MAJOR CROPS

Major crops refer to the sixteen or more plants that usually comprise 80 percent or more of the total crop composition for a given tablón. A majority of the tablones in Panajachel were surveyed for crop character, but the quantitative data on precise crop character is drawn from the transect. The transect is generally representative of a cultural cross-section of the delta as a whole, although Tax (1953:45) pointed out that "particular families and 'race' groups usually follow consistent and commonly known patterns." The transect involved the west side of the delta. Therefore, cropping patterns followed by native Panajacheleños (the east side being their "stronghold") are possibly underrepresented. Comparing data taken in both sectors suggests that microregional differences between the two districts has lessened since the 1930s.

Onions

Onions (Allium ascalonicum L.) continue to be the predominant crop grown on Panajachel's tablones. In late November and early December, onions were the predominant crop in 29 percent of the active tablones. Portions of 9 percent of the tablones were being used for onion seedling nurture. In another 3 percent of the tablones a "condominium" relationship existed between onions and another major crop. Therefore, roughly 40 percent of the tablones were committed to onions. Tax (1953:44) said that 45.8 percent of the tablón acreage was used for onions during the same time of year.

Several things could account for this difference. Cropping practices have changed during the last three decades. Even if slightly less land is being used for onions, an intensification of methods has occurred that has allowed the farmer to increase production. Closer spacing of individual plants and more frequent transplantings are part of the strategy of intensification.

There are fluctuations from year to year in the amount of land committed to onions. It was suggested to this author that the price of onions varies enough from year to year, so that in some years it is scarcely worth the effort to grow them.

Onions are found in a variety of contexts. At one extreme, onions are monocropped, occurring in about 10.9 percent of the tablones. The remaining onion tablones have many different patterns of intercropping. All possible crop combinations involving onions and the other 150-odd tablón plants probably have been tried in the past.

There may be a positive correlation between the use of fertilizers, degree of acculturation, and "monocropping." Where farmers tended to be more fastidious in limiting the variety of crops per tablón, more chemical fertilizer was used. Monocropping may have been part of the instructional baggage that accompanied the introduction of chemical fertilizers by the Servicio de Fomento de la Economía Indigena (SFEI) in 1959.

Garlic

Like onion growing, garlic (Allium sativum L.) production symbolizes and distinguishes Panajachel's tablón culture. Garlic occupies 11 percent of the tablones with single crop predominance. This figure is the same as Tax gives for his total yearly average, but falls 8 percent short of the figure Tax cites for the month of December. Currently, the onion/garlic plantings may be delayed until the middle of December or later.

Many of the tablones used for corn in the wet season are succeeded by garlic in the dry season. Farmers who still had corn in the field during the first week in December said that they would replace the corn with garlic after the harvest. This was one of the few cases in which clear statements were offered concerning the exact sequence of crop successions.

Strawberries

Strawberries (Fragaria chiloensis) represent the biggest change in the cash cropping pattern since Tax's survey. According to local recollections, they first appeared in the Panajachel tablón crop complex during this century. Manuel Crespo, local entrepreneur and long time resident, dates the introduction of strawberries from the 1930s. He suggested that Gordon Smith, an English planter and agronomist, first attempted strawberry raising in his experimental garden plots in Panajachel. Juan Maria Sallas, one of Smith's assistants, introduced the practice to the local farmers. The spread of strawberry raising must have been slow during this time. Tax makes only passing mention of their cultivation.

Today, strawberries have become the fifth most common tablón crop on the delta, aided by the introduction of chemical fertilizer. They do not occur on the hillside tablones, suggesting either cultural or environmental proscriptions, or both. Very recently strawberries have declined in popularity as a preferred cash crop, which may be a result of the rising fertilizer prices. This problem was pointed out on several occasions by farmers complaining of the high fertilizer costs incurred during the latest survey period.

Strawberries make heavy demands on the soil. They are grown most successfully on the "new lands" beyond the immediate flood plain. In 1964 Hinshaw (1975:190) was able to say: "They had become possibly the second most important truck crop." Hinshaw did not have the necessary information to say whether this was in terms of land occupied or cash returns. Today, they occupy about 4 percent of the tablones. Total acreage devoted to strawberries may be higher, as the average tablón plot used for strawberries varies considerably in shape and size. The strawberry tablón is a distinct morphological subclass of tablón, in which width often equals length. Therefore, as much as 10 percent of the total tablón acreage may be planted in strawberries, but this still places them in fourth or fifth position behind onions, corn, garlic, and beans in overall spatial importance.

In terms of change and innovation, strawberries represent the rise and fall of a succession of "risk" crops in the delta. In microcosm, this mirrors the characteristic tropical boom and bust patterns generated by searches for vegetal produit moteurs, to use Pierre Chaunu's expression. Strawberries are the probable successors of the melon pear, or pepino, which was the principal "risk" crop of the time of Tax's studies. The decline of the pepino is discussed below.

Beside demanding exogenous fertilizer (either chemical or animal), strawberries create three more problems. First, they are land extensive, requiring more space than onions or garlic. Therefore, as land becomes more expensive and less available on the delta, strawberry production may be cut back. Secondly, they are almost exclusively produced for the urban Ladino market, and are subject to the vicissitudes of its demand. Finally, strawberries must be quickly moved to market, and they have cosmetic standards that the Indians do not always observe. Given these limitations, the "strawberry boom" is on the wane. This will probably continue unless the market grows more rapidly than at present.

Beans

Beans (Phaseolus spp.), both "green" and the variegated selection of "dry" pulses, occupy an important place in the tablón complex. Beans have risen in price some 10 to 15 fold since 1936, with much of this increase coming about in the last few years. This is higher than the average rise in prices for most other tablón crops and has been in response to the rising worldwide protein demand. Perhaps as a consequence, delta farmers are committing about 10 percent of their land to beans, up from 8 percent in 1936. This figure is probably higher in actuality, because Tax reported no beans at all for the month of November and only 0.7 percent for December. It might be inferred that the annual figure for Panajachel today is closer to 15 percent, or twice as much as formerly raised. In addition to the dried beans' role as a nutritional staple, beans are seen as a form of security in the face of the perplexing rise in prices. Once content to raise only limited amounts of beans, and buy the rest later during the year, the farmer is apparently drifting toward more self-sufficiency in this regard (Plate 4).

Green beans are grown mostly on poles of caña veral (Gynerium spp.) one-half inch thick and five or six feet high, spaced close together--the average distance is 20 cm between poles. While they are distributed evenly over tablones--space economic considerations are paramount--they are not grown in rows per se. Pole beans are grown in greatest numbers on the steep slopes above the delta. Here, pole beans are grown on narrow, terraced tablones (3 m wide or less) that follow the contours of the slopes, negating the horizontal space limitations.

Two varieties of ground beans (Phaseolus vulgaris)--white and black--are grown. Like onions, garlic, and strawberries, they are grown either alone or intercropped with innumerable combinations of other useful plants. Not much use was noted of the standard Mesoamerican intercropping scheme combining maize and beans. Panajacheleños prefer to grow pole and ground beans separately from maize, although the majority of beans are grown in the wet season, as is maize.

Maize

Maize (Zea Mays L.) remains the second most important crop grown in tablones, and if one includes the maize grown in shifting milpas outside the delta by Panajacheleños, then it is the primary crop of the village. Maize is so much taken for granted throughout Guatemala that Carter (1969:14) reported upon questioning the Kekchi around Lake Izabal who failed to mention maize as an item in their diet! While maize does not exercise the "dietary hegemony" for the Panajacheleño that it assumes for the lowland shifting cultivator, it is still the basis of subsistence.

Figures on how much, if any, of the maize grown in tablones is marketed could not be calculated, due to the timing of the study. Probably it is negligible. In 1936 the adult Panajacheleño ate between 200 and 220 kg of corn per year (Tax 1953:165). This is slightly under the 0.5-1.0 kg-per-day figure generally cited for shifting Maya cultivators (Benedict and Steggerda 1937; Morley 1946; Higbee 1948; Stadelman 1940). The Panajacheleño diet is probably more varied than that of the Mayan farmers practicing extensive maize cultivation. Present figures are surprisingly low for maize production on tablones without aid of chemical fertilizer. Yields of only 55 kg per cuerda of tablones (.07 hectares) were reported. Therefore, in a year with average rainfall (maize is not normally irrigated in tablones), the yield would be 786 kg per hectare. This equals second- or third-year yields under shifting cultivation in many parts of Guatemala. The difference is that the same land is cultivated year-in and year-out in the tablón complex. Taken over a 24-year period (counting dry years in which yields drop to as low as 400 kg per hectare), one might expect to harvest between 12 and 18 tons of maize per hectare. This compares with yields of 18 tons per hectare (over 24 years with a 1:3 swidden cycle) for the "enormously productive" swidden cultivation of the Grijalva flood plain in Tabasco (Sanders 1957:312).

In sum, the yields from the delta of Panajachel are probably equivalent to the yields from shifting milpas on the neighboring mountain slopes, but not nearly enough maize can be grown on tablones under current practices (non-irrigated, not chemically fertilized) to support familial needs. Therefore, only one quarter of one person's yearly maize requirement can come from a cuerda of tablones. Moreover, the average size landholding along the transect is between two and three cuerdas per family. Thus few families can grow enough maize from their tablones to support more than one member for the year. Clearly, their maize comes from elsewhere.

Much of the maize comes from milpas owned or rented on the hillslopes above Panajachel or in and around neighboring villages.[1] One informant, a transplanted Jorgeño living in Panajachel, maintained ties with the communal milpa lands owned by San Jorge, which are some distance from the village on the western slopes of the Quiscab Valley. He made weekly trips of some 30 km to work communally in the milpa. To this extent, the residents of Panajachel participate directly in the exploitation of a wide range of ecological zones. Similar arrangements for utilizing the vertical diversity of tropical regions for subsistence ends has been described by Brush (1973).

Many plants are intercropped with corn. The idealized maize/pulse/cucurbit triad is not dominant in Panajachel. The "ranks of trinity" have been opened, at least in Panajachel, to allow güisquil, cintula, manioc, sweet potatoes, chipilín, malanga, hierba buena, and many other cultigens to grow under, over, around, and beside the mounded stalks of corn.

Carrots and other Old World Vegetables

Carrots (Daucus carota), radishes (Raphanus sativus), lettuce (Lactuca sativa), beets (Beta vulgaris), turnips (Brassica rapa), and cabbage (Brassica oleracea) are all found in tablones in significant numbers. These six European domesticates play a lesser role than the five major crops, however they dominate the crop composition of tablones in scattered cases. Of this group, only carrots approach the 1 percent level of frequency. All six are widespread as intercropped companions of onions, garlic, strawberries, and beans.

Lettuce plants appear as isolated individuals in many onion and garlic tablones, suggesting adventive origins. In this regard they are similar to adventive native greens and pot herbs. The other five vegetables are sometimes nutured in seed beds. Like onions, they are transplanted once, twice, or even three times to maximize spatial economy. Cabbage is grown along the tablón sides near the rims, or "ears." In this location it fills a space on the tablón sides above sweet potatoes, arracacha, and peanuts. Root crops occupy the middle section of the tablón sides.

Seed beds, or semilleros, for these vegetables are made with piled rocks removed from the soil and arranged in square or rectangular mounds about 1 m high (Plate 5) which contains specially collected soil. One or more vegetables are raised from seed to be transplanted in tablones. The location of the planters further illustrates the spatial efficiency of the system. The planters are placed at the end of tablones or built around rock outcroppings too large to be removed from the fields. They may serve an additional function. As Nutall (1920) pointed out, rocks left in fields may act as heat foils, regulating microclimatic conditions: "Native gardeners had learned through long experience that many plants do better among rocks. Rocks (left in the field) not only conserve moisture, but also conserve solar heat to counteract the cold nights at these altitudes." Kropotkin (1913:125) also pointed out that rock walls and large rocks (left in fields) play an important role in regulating heat in horticultural plots.

Because it is difficult to estimate the total quantities yielded, it was difficult to calculate the economic returns from some of these vegetables. Much of the scattered plantings of radishes, turnips, and lettuce are consumed directly by the farmer. Beets and carrots are cropped with more frequency and in larger quantities than the others. Accordingly, they show up more in the local market. Therefore, production exceeds direct consumption.

The situation with cabbage is different and more nearly approaches the status of a cash crop. Tax (1953:114) remarked that of the vegetables grown from imported seed, "only cabbage gives promise of much profit--and correspondingly it is risky." Most of the cabbage sold in the Sololá regional market comes from the tierra fría. Little comes from Panajachel. The delta is located toward the lower (warmer) limit of the cabbage range. Panajacheleños can grow other vegetables more profitably.

ROOTS AND TUBERS

There is debate concerning the importance of root crops in studies of Maya subsistence patterns, both ancient and modern. Bronson (1966) argued that manioc provided the subsistence base of the lowland Ancient Maya civilization. Manioc and other roots and tubers may have rivaled the dietary dominance of maize in parts of the Maya realm. Root crops are unquestionably important in Panajachel today.

About 32 percent of the tablones have either manioc, sweet potatoes, or arracacha. Many have all three. Tax estimated that about 25 percent of the tablones grew root crops as an adjunct to vegetables or maize. The current figure is around 30 percent. In 1936, of the families that Tax surveyed for food consumption, the net weight of root crops intake (excluding potatoes) compared to maize and lime consumption, ranged in ratios from 1:4 to 1:98 of total intake. If one includes other proficient sources of carbohydrates (sugar, potatoes, beans, and squash) to the 1936 totals, then only 10 percent (and probably much less) of the carbohydrate consumption was supplied through root crops. Though no firm data could be found, it appeared that, based on personal observation, this has changed.

Most informants said that they ate sweet potatoes as a daily routine. Manioc consumption was less frequent, but still a common food item. This suggests that root crops will become increasingly common in tablones and serve as dietary substitutes for the more traditional staples: maize, bean, and meat. It might seem reasonable that maize and bean production would be increased in an effort to counter the swiftly rising prices for these items. However, both maize and beans are relatively space inefficient. To expand their production would involve capital outlays in terms of new tablones built and new space acquired. Instead, Panajacheleños are intensifying root-crop production, along with a greater effort at producing "space economic" cash crops like onions and garlic.

Manioc

"Sweet" manioc (Manihot esculenta) is the main type grown in the Lake basin. The origin of manioc in the tablón complexes is uncertain. Tax (1953) suggested that local tradition dates the introduction of manioc from around the turn of this century.[2] If this is true, it may have been a result of the forced migrations of highlanders to the lowlands at the end of the period of mandamientos. During this era, transhumant highland Indians fed themselves while working on the lowland plantations. Manioc culture would have been a logical and relatively undemanding means of self-provisionment. In this sense, manioc, sweet potatoes, and arracacha (tropical lowland root crops) symbolize the intrusion of Outer Guatemala into Inner Guatemala. Raising root crops represents a dietary strategy that can be turned to in times of hardship or external pressure and change.

Because it may represent manioc's highest altitudinal extention anywhere in the tropics, manioc cultivation in the Atitlán basin may be an anomaly. Its appearance at Panajachel stretches the limits for Central America. Cowgill (1971:51) gives the upper range as: "500 to 700 m when other conditions are favorable." It is found growing at 1,600 m in Panajachel, or 900 m higher than Cowgill's stated range.

Again, this points out that Panajachel is blessed with rather remarkable conditions for growing a wide range of crops. These conditions include its confluence at the conjuncture of several climatic zones, a sheltered north-shore position on a sizeable body of water, and its position at the "cultural interface" between the tierra fría and the tierra caliente.

Manioc is grown individually, in disparate clumps, or in rows along the top edge of the tablón. It is generally allowed to grow for two years before harvesting. Therefore manioc represents, like tree crops, semi-permanent features in tablones. They are not dug up each time the tablón is reworked. Moreover, manioc can be left in the ground, creating an in situ storage arrangement that allows the farmer to keep some of their "starch capital" unharvested.

Sweet Potatoes and Arracacha

Both sweet potatoes (Ipomoea Batatos (L.) Pax) and arracacha (Arracacia xanthorrhiza) are grown throughout the delta, though sweet potatoes are far more common than arracacha. Most often these root crops are grown along the sides of tablones and occupy a niche that might otherwise go unfilled (Plate 6). At the same time, they exploit moisture in the irrigation ditches that might go unused. These crops serve an additional function when planted on the sides: they help stabilize the tablón walls.

Sweet potatoes are the most universal of the three lowland tubers occurring in Panajachel. Like manioc, they either represent a recent introduction from the coastal lowlands, or equally plausible, may have been an important part of the crop complex that supported large pre-Conquest Mayan populations in tierra templada areas, like the Lake Atitlán basin.

"Irish" Potato

The common potato (<u>Solanum</u> tuberosum <u>L</u>.) is a relative rarity in the delta tablones of Panajachel. The mesothermal conditions of the valley floor create a marginal environment for the potato, appearing to be at the lower limit of its range. Several isolated plantings were found (at about 1,610 m) in the northern most tablones of the transect. Therefore, there may be distinct crop ranges within the flood plain. McBryde (1947:31) felt that "it is too warm for potatoes at Panajachel." Tax makes no mention of potato raising on the delta.

Cultural factors may also play a role. The farmer growing potatoes at the north end of the transect is a renter from one of the dispersed settlements near Concepción. Hailing from a tierra fría community, the farmer may be experimenting with a familiar crop at an uncommonly low elevation. This supports Tax's (1953) assertion that the farmers of the Panajachel delta are willing experimenters.[3]

Hierba Mora

Other tablón-grown nightshades or members of the family <u>Solanaceae</u> have been discussed above. An additional nightshade, hierba mora (<u>Solananum nigram</u>), is fairly common, though not a central item in the tablón system. It is representative of numerous other herbs, spices, condiments, and greens, both indigenous and introduced, occurring throughout the system. These disparate plants are important as medicinals, condiments and dietary supplements.

OTHER MAJOR CROPS

The following crops are being included in this study, even though each appear only once or twice in the transect survey in numbers large enough to dominate the character of a given tablón.

Melon Pear

The melon pear (<u>Solanum Guatemalense</u> hort.; <u>S</u>. <u>muricatum</u>), or <u>pepino</u>, has been mentioned in reference to strawberries. Both represent "risk" crops that have enjoyed a period of profitable florescence followed by decline. Tentatively, the decline can be associated with soil exhaustion. The melon pear is probably a recent introduction to Guatemala, perhaps from South America (McBryde 1947:141).

McBryde thought that Panajachel was the sole center of pepino raising in southwest Guatemala. In 1936 pepinos, among the five major crops grown in the delta, was the premier crop in terms of economic returns. Considerable brujería seems to have been connected with their culture. Tax (1953:131) reported that pepinos were introduced from San Antonio within the memory of the informant. They soon became a specialty of Panajachel. At the same time, some growers were apprehensive that they may have stolen the "spirit" of pepino culture away from the Antoñeros. They feared that retribution would come in the form of losing their onion monopoly to these neighbors.

Another aspect of ritual accompanies pepino production and arises from the care they demand. They are not grown on tablones per se, rather they are grown in small mounds that look like rows of miniature volcanic cones (Plate 7). The mounds are about 30 cm high and 75 cm wide at the base. The pepinos are grown inside the cone.

During 1974, only one farmer grew pepinos in the transect sample. None of the locations that McBryde (1974:122) gives in his map are areas where pepinos are grown today. During that time (the 1930s) the areas listed by McBryde were relatively "new lands." These were on the west side, and formerly had been a large cane break opened up by Jorgeños in the previous two decades. This corresponds with Tax's opinion that pepinos had to be grown in relatively fresh soil, or yields would drop off precipitously after the first year. Paradoxically, the major concentration of pepinos today is in the center of Jucanyá, presumably the "oldest" cultivated land in the delta. From distribution alone, one might conclude that this plant has suffered a considerable decline in importance.

Peas

Sweet peas (Pistum sativum) are another introduction that has declined in importance. According to this author's informants and to Tax (1953:130), peas were once an important crop "grown by the cuerda in garden beds, or like corn." Other villages have taken up their cultivation and are able to grow them with more success. It may be that peas are not as remunerative as some other crops. As suggested by one fieldhand, the space and skill available in Panajachel can be put to more rewarding ends. This suggests that there is a definite succession of crop types occurring in Panajachel. As lands become scarcer, space demanding plants like peas, tomatoes, and squashes are replaced by more "space economic" plants like onions.

Brambles

Several strains of brambles (Rubus spp.)--raspberries, blackberries, and the hybrids thereof--occupy some tablones on the east side of the delta. They seem to be grown at the instigation of chalet owners who rent out adjoining land for tablón farming. Like strawberries, they were probably introduced during this century by Gordon Smith or some other experimentally minded planter. Their potential as a profitable cash crop within the tablón context seems better than strawberries, given their vertical orientation. More blackberries or its allies can be grown per cuerda of tablón land than strawberries. The problem (Why more berries of this type are not grown?) seems to be one of demand. Rubus berries are not part of the local diet. If Panajachel continues to grow as a tourist and a "second-home" center, then a market for Rubus berries will be close at hand.

Coffee

At first glance, including coffee (Coffea arabica var.) in a discussion of tablón crops appears contradictory. One of the critical constructs of Tax's argument in Penny Capitalism is the motive tension created between the demands of land for coffee production (chiefly by Ladinos) and the need for truck land (by Indians) for vegetable growing. While recognizing and agreeing with his analysis that caste and class divisions can be seen symbolically in this competition for land use, the fact remains that some coffee growers start their coffee seedlings in tablones.

The English planter, John Vickers, might be responsible for this practice. In the 1930s one traveler described this locally famous figure as "one Englishman... that had 'gone native'" (Halle 1936:160). Without further research it is impossible to say if, or just how far, John Vickers had "gone native"--perhaps enough to adopt the Indian's tablón techniques into his coffee planting strategy. His son, Edward Vickers, maintains some twenty or more tablones for this purpose. Some Indians do the same on a smaller scale to provide coffee seedlings for their house gardens.

ABORIGINAL SEED PLANTS

Amaranths and Chenopods

Throughout much of indigenous America, cultivated amaranths (Amaranthus spp.) and chenopods (Chenopodium spp.) are survivals of formerly important food plants. In most areas they have been eclipsed by grains, like maize and wheat, or by roots and tubers as the bases of subsistence (Sauer 1936). Two species of amaranths are cultivated in tablones. Amaranthus paniculatum L. is known locally as bledo colorado, while A. caudatus is the less common cola de zorro. Together the two amaranths are known generically as bledo, and are found in about 10 percent of the tablones.

The common chenopod is known variously as apazote, epazote, or hierba pazotemente. Both bledo and apazote are fed to fowl, and cooked with beans for meals. Combining pulses with other plants grown for their edible seeds (maize, amaranths, sunflowers, etc.) potentiates protein synthesis when eaten by humans. If Sauer was correct in positing an "Ur-complex" centered on amaranths precursor to the maize/bean/squash triad, this combination of beans and amaranth in Panajachel may be a survival of an old practice.

At the very least, amaranths and chenopods are weedy "camp followers" in the tablón world, afforded and maintaining a low profile in the total picture. Nonetheless, bledo is found in 9 percent of the tablones, and apazote in about 4 percent. Their greatest concentrations are in tablones left fallow for six months to a year. Therefore, their presence is doubly deceptive. Finally, most informants took the bledos for granted when asked to point out all economically useful plants in the tablón before them.[4] To this student, it seemed to be a clear case of extreme familiarity, rather than disutility, that prompted the failures to mention bledo and apazote.

A final note is of interest here. The Jorgeños of the west side (tablones 1099-1249) had no apazote and less than 10 bledo plants in their tablones. They are among the most conscientious monocroppers on the delta. Also, they were the only users of herbicides and pesticides that were observed during this field work. More study on the degree of "modernity" in cropping practices and the variety and antiquity of crops grown needs to be done before one can suggest that there is a positive correlation between these two variables. Tentative observations suggest that there is such a relationship, pointing to a cultural/historical research problem of some interest.

Squashes

Because Panajachel is relatively close to the place of origin of many cucurbits, and its farmers are receptive to growing many different crop plants, one might expect that a number of squashes (Cucurbita spp.) would be grown. This is not the case. The squashes C. moschata and C. pepo, locally known as ayote and güicoy, are infrequently grown in tablones. Much less than 1 percent of the tablones of the sample hosted these or allied cucurbits. The few tablones that did were also planted in maize. Moreover, little squash is grown in the hill milpas, if Tax's (1953:124) monitoring of produce entering the local market is indicative. No squash was reported for several days in the months of March and April during 1937. Tax's data for squash production in general was "incomplete," but it suggests that squash was sparse, even in the milpa precincts where maize and beans dominate the crop character. This suggests that Panajacheleños have relied on other elements to complement the traditional "triad."

Güisquil (Sechium edule), though not a member of the genus Cucurbita, is a squash-like plant found in and around tablones. The "vegetable pear," or chayote as it is called in Mexico, is a popular climbing plant that exploits fences, outbuildings in the fields, and rock piles or outcroppings in or next to tablones. It is a common item in the Panajachel market. In part it fills the void in the tablón complex created by the absence of Cucurbita spp. proper. Zhietnev felt that güisquil may have been spread with the Toltec cultural advance into Guatemala (McBryde 1947:137). While it is widespread in Mesoamerica, and probably not specifically adapted to tablón conditions, it does have a vertical orientation that allows it to grow in tablones with greater space economy than any of the other cucurbits. One could make an inference in this regard: if the Toltecs introduced or popularized güisquil (and "displaced" the other cucurbits in the process), it may have been in concert with tablones or raised garden beds of a similar nature.

LEGUMES

Chipilín

Even though its role as a legume may be quite important in the total tablón system, little study of chipilín (Crotalaria longirostrata) has been done. McBryde (1947:143) mentions it only in passing, as does Tax (1953:173). Yet chipilín is found in 19 percent of the tablones studied, which places it in the top quartile in terms of crop species frequency. Often the leaves and flowers are added to tortillas and soups as a soporific or sedative. According to U.S. Peace Corps workers in the area, chipilín is also an important source of iron and other minerals for many Indians.

In addition to providing the Indians with minerals and a mild narco-stupefacient, chipilín is a leguminous shrub. It is undoubtedly important as a supplier of nitrogen to the soil. This aspect may be a crucial one. From the data collected, it appears that chipilín is one of the first native crop plants abandoned as a farmer moves toward intensive chemical fertilization and monocropping. It may be that rotating other legumes (beans) in tablones obviates the need for chipilín. In highland New Guinea, the casuarina tree serves an important function as a nitrogen "fixer" within native horticultural systems (Clarke 1971). Chipilín probably serves a closely analogous function in Panajachel's horticultural system.

Finally, chipilín is grown widely on tablones at Panajachel, but not grown on tablones at higher elevations. Chipilín grows wild at higher elevations and is gathered by Jorgeños and others; therefore, altitudinal limitations do not seem to be a problem. This seems to be another situation in which specialization by village delimits the range in which a crop is found.

Peanuts

The other two legumes found in the tablón system are beans (Phaseolus spp.), described above, and peanuts (Arachis hypogaea). Beans are important in terms of nitrogen fixation to the system as a whole. The Indians say that rotating beans in the tablones "gives force"--"da fuerza"--to the soil. This may explain why tablones are used for bean culture. Beans could be grown just as easily in shifting fields on the mountain aides.

Peanuts, relatively rare in the tablón system, are grown on the sides of the tablones interspersed between sweet potato cuttings. While peanuts are a popular treat for festival or market days, daily consumption of peanuts is minimal.

NIGHTSHADES

Chiles

Several varieties of chiles (Capsicum spp.) are grown in tablones. They are usually grown as solitary individuals, though clumps or vague rows of chile plants are sometimes cultivated. Varieties are distinguished by their coloration and uses, with the black variety being the most common. Chiles seem to be raised like many of the native crop plants--as items for home consumption rather than objects for market. Chiles are always planted with other crops, usually with onions or a broad range of plants grown on tablones--the exception being corn.

Tomatoes

There are at least four or five different tomatoes (Lycopersicon spp.; Physalis spp.) grown involving two genera. The improved varieties of Lycopersicon are occasionally found. Apparently these European "introductions" are too sprawling (much like the squashes), and their foliage too dense to be of much interest to tablón cultivators. One of the principals of tablón culture seems to be: Any plant demanding too much horizontal space relative to the size of its fruit will be culturally selected against. Therefore, the feral, or retrograde tomatillo (tomate de culebra) (L. esculentum var. cerasiforme) is much more popular. This "wild" tomato grows almost like a bush, with physiognomic resemblance to the chile plants. In this form it does not monopolize a wide space, nor does it block solar radiation from the vegetables growing under it.

Even more common than the tomatillo is the miltomate (Physalis spp.). It is also known as the groundcherry (not to be confused with the cherry tomato (Lycopersicon). The ground-cherry is a weedy bush that bears a greenish fruit used in sauces and soups. It has a durable husk, so it can be transport-ed much more easily than varieties of Lycopersicon. It shares a space-saving vertical orientation with the tomatillo. The miltomate reaches 1 m in height, therefore a large number of "understory" or surface crops such as onions, lettuce, and beans can be grown directly beneath it.

TREE CROPS

Judging from written accounts, the image of Panajachel in the colonial mind involved a site profusely planted with fruit-bearing trees of many families, both native and introduced. This Edenic image, originally promoted by the Franciscans settling in Panajachel, is accurate today if the range of fruit-bearing trees are inventoried. Throughout the delta, citrus, palm, avocado, nut, and many native fruit trees provide their owners with suste-nance and cash that involves little effort. Panajachel may well be the "Spanish plum" or jocote (Spondias spp.) capital of the world, if diversity of forms is the criterion for such a distinction. As noted earlier, many citrus as well as native fruit trees find optimum conditions at Panajachel.

Fruit Trees

The primary reserve of fruit trees is the housegarden com-pound. Housegarden compounds include nearly all available ag-ricultural land on the flood plain that is not used for cafetales, tablones, milpa, or pasturage. Many of the seedlings of cultivated trees are started in tablones, and then later transplanted to housegardens. There are several advantages. First, the seedlings are looked after. They are safe from the traffic of pets and small children that play in the house compounds. Second, in tablones the seedlings receive the benefit of fertilizing, either by the or-ganic or chemical treatments applied to tablones. Third, in tab-lones the seedlings are watered during the dry season. Rarely does the number of trees (including papayas) exceed three in a single tablón. The problem of shading may be responsible for lim-iting the number of trees allowed to reach maturity in tablones, but the probable explanation is that a tree (or shrubs like garbanzos, chiles, chipilín, manioc) present a nuisance each time the tablón is reworked, which may be as many as four times a year.

Other Trees

Trees raised for purposes other than fruit production are also grown in tablones, but rarely to maturity. The major function of these trees is twofold. First, they provide firewood. (Similarly, manioc branches are used as firewood for preparing meals in the fields). Second, they are grown as canopy covers for coffee trees. Panajachel's most frequent "madres de cafe" are ilamo (Almus acuminata) and gravilea or "silk oak" (Grevillea robusta).

Trees found in tablones are often volunteers, started from seeds washed down the irrigation ditches and left to grow and later to be transplanted. Therefore, there is no conscious plan for planting many of the trees found in tablones. In this sense, trees are analogous to many of the weedy herbs and greens found in the system.

Willow

Sauce, or willow (Salix chilensis), are used for boundary demarcation and hedging. In one case, sauce is found in a group of chinampa-like tablones (Plate 8). This situation involves several raised garden plots, resembling chinampas more than tablones, which are located near the center of town. The plots were once a swampy area that has been reclaimed for vegetable farming. Presumably the willows are used for transpirative purposes (for water level regulation?) and for tablón anchorage, though the latter use seems hardly necessary today.

Springs at one end of the "chinampa-tablón" plots allow water to stand throughout the year. This was demonstrated by the fishing and frogging successes of small boys at the site. The plots are rented out by the owner at almost four times the price of a similar cuerda of land near the river. The renter claimed the crop returns were correspondingly high. Fertilizer for these plots is dredged from the irrigation ditches, much like the descriptions given for chinampas. L. C. Stuart, the noted zoologist (pers. com.), felt that they had been constructed in part during his memory (since the late 1920s). However, when questioned, the local people say that the plots have been there "always" ("para siempre"), like many features of the cultural landscape. These "chinampas" may well represent pre-Columbian relics dating back to the Toltec invasions. To the knowledge of this author, nothing similar has been reported elsewhere in Guatemala.

Prickly Pear

Certain species of prickly pear (Nopalea and Opuntia) are cultivated for their fruit, known as tuna, and eaten by Indians. These cacti are grown in the rock rubble piles that accumulate in lattice-like patterns around tablón plots where rock removal is a problem. Before the 1860s, several species of tuna plants enjoyed economic importance as hosts to the cochineal insect (Coccus cacti). Used since antiquity for a red dye, the cochineal industry was destroyed by the development of aniline dyes. Like the cucurbitoid güisquil, many of the Nopalea species were introduced from Mexico during the pre-Columbian incursions. Unlike güisquil, they are not an adaptively significant addition to tablón culture. Today, they are only an occasional item in the tablón complex.

Pitahaya

Pitahaya (Hylocereus undatus), sometimes called the night-blooming cereus, is a cactus that inhabits tablón-associated rock piles. It bears a bright scarlet, black-seeded fruit that is peddled on the local beach, but is seldom found in the market.

Banana

Bananas (Musa spp.) are often grown in clumps or individually near tablones, but not on tablones. The shade produced by bananas, and probably the excessive water requirements, prevents planting them within tablones. Moreover, the root structure (surficial) would be an obstacle to remaking the tablón. Most of the bananas occur on the sides of the steep 2 m high benches created by former floods in the lower west tablón area. Other bananas are planted along the courses of major irrigation channels. The fronds provide material for building shelters within the tablón precincts, although it is increasingly more common to see metal or plastic used for roofing in these situations.

ORNAMENTALS

Some twenty or thirty species of flowers are grown in tablones. The majority of these flowers are native to the New World. As previously suggested, one explanation of tablón function (if tablones are pre-Columbian artifacts) concerns their possible role as ceremonial garden plots.[5] Again, it can be suggested that tablones were introduced or "invented" as early as the Toltec theocratic invasions. Numerous accounts from the Conquest chronicles mention the importance of flowers as ceremonial objects. Today, flowers are an important aspect of the cultural life of Panajacheleños.

Flowers are dispersed in tablones throughout the delta. Like the larger spice and herb plants, they usually appear as isolated individuals. The exception is azucena (Lilium candidum), which is intercropped with strawberries in considerable numbers, or, in some cases, is the predominant crop. Azucena is grown in seed beds similar to the beds used for starting onions and beets. A number of different flowers are planted along the upper walls of tablones, adding to the aesthetic appeal of these artful structures. In this role, flowers may serve the function of adding yet more beauty to a segment of landscape in which many Panajacheleños spend most of their waking hours. Seen in this way, tablón culture is the agricultural analogue of the area's textile craft. Both tablones and textiles are meticulous creations, aesthetic to the verge of excess and labor intensive to a degree equaling any of their cognate activities in indigenous Mesoamerica.

OTHER TABLON PLANTS

Other than the plants already mentioned, perhaps as many as 150 other species are cultivated or tended in tablones. Several plants grown in close association with tablones should be mentioned. These plants do not, under ordinary circumstances, grow in tablones, but grow around them. Two plants are commonly used as mojones, or live boundary markers. The most common is yucca (Yucca elephantipes), sometimes called hizote. Deep rooted, it grows for decades. In early times it was probably quite important as a flood-plain boundary marker that could withstand flooding. In the wake of Hurricane Fifi (November 1974), this author observed hizote plants nearly buried by rubble that destroyed all traces of former tablones, yet the farmers were able to use these markers in the succeeding weeks to reestablish their boundaries. The leaves are universally used as fiber to bind bunches of onions. Standley (1930:228) suggested that the yucca was originally brought from Vera Cruz, adding a bit of inferential fuel to the proposition that the Toltecs introduced tablón culture to the Guatemala highlands (Plate 9).

NOTES

[1] The Jorgeños have greater reserves of communal land for growing milpa than the Panajacheleños, whose farming activities are largely confined to the delta. This may explain the proclivity of the Jorgeños to monocrop onions and garlic more than Panajacheleños. Jorgeños have a wider range of "vertical zonation" to draw upon. They can ignore the root crop and peripheral pot herb production that characterizes the Panajacheleños tablones, and concentrate on cash cropping instead.

Plate 1.   Tools used at Aguacatán for raising garlic.

Two widths of Aguacatanian hoe shown; paddle is for forming sides of tablón. Lagenaria gourd at bottom is used for splash irrigation. Other containers hold garlic cloves.

Plate 2.   Ranchito used for holding costumbres.

Incense burners are hanging on the left side wall. Tablón 1127 is in the immediate foreground. Black beans and manioc are being grown. Rocks are being used to rim the relatively low tablón sides.

Plate 3.   Fishing canoe and harvested reeds.

Canoe is made from an avocado log. The tule reed is an important refuge for resident and migrating water fowl.

Plate 4.   Tablón with black beans.

Nearly mature onions appear in the adjacent tablones.

Plate 5.   Seed beds for cabbage and other vegetables.

Semilleros, or elevated seed beds, serve several functions. They help solve the problem of rock removal, they act as insulation, and they protect the seedlings from some pests, such as pocket gophers. Note section of western main wall beyond planters and terraced tablones on hill slope.

Plate 6.   Sweet potatoes and arracacha planted on tablón sides.

Radishes, beets, and turnips are intercropped with onions on tablón tops.

Plate 7.   Melon pear mounds.
Sweet potatoes, manioc, flowers, and peach seedlings intercropped on this melon pear tablón.

Plate 8.   Willow trees and chinampa-like tablones.
Chinampa tops have been cleared of rainy-season maize crop and await dry-season planting of garlic.

Plate 9. Yucca plants.
Top of tablón is covered with pine needles to protect onion seedlings beneath.

Plate 10. Removing rocks by sifting with screen to make new tablón land.
Reclaiming land by this method requires considerable labor. In this case, five male members of one family had worked ten days to create 45 square meters of "soil" (mostly sand). Therefore, some 800 man-hours are required to construct a 90 square meter tablón. This does not include time (usually over many years) needed to improve soil through green manuring. For this reason, chemical fertilizers are popular.

Plate 11. Forming sides of tablón.
After clearing and cleaning, wetting the soil with water is the initial step.
Tablón being worked is number 1054.

Plate 12. Forming sides of tablón.
The second step includes piling dirt for the "ears," or rims. An azadón is used.
Sharpened cane spikes in Tablón 1053 are used to discourage dog traffic.

Plate 13. Forming sides of tablón.

A final step in making the tablón sides. Farmer uses azadón blade as a tamper.
Wooden box is used to transport topsoil to "appreciate" soil capital. This soil is
often "borrowed" from coffee fincas at some distance from the delta.

Plate 14. Banana frond conduits for irrigating terraced tablones.

Pole beans are being grown on narrow terraced tablones, located north of Media
Cuesta above 1,700 m.

Plate 15. Chinampa-tablón tomas being cleared.
Accumulated muck from the tomas is applied to tablón tops as fertilizer. Water level is normally maintained within 30 cm of tablón top, but drained for tablón-remaking twice a year. Fish and frogs are collected as water is drained.

Plate 16. View of delta from Media Cuesta.
Lower right corner of plate shows tablones on Media Cuesta at about 1,720 m. Note the average width (12 m) is much greater than the width of tablones on the flood plain.

[2] While manioc may have been introduced into Panajachel during the last century, as Tax has suggested, chroniclers mention manioc at various sites within the basin during the sixteenth century (Paez-Betancor 1964:103). Moreover, Diego de Ocana (1933:301) mentioned manioc in connection with the expansion of the irrigation systems during this period in the basin: "...and were and more, almost all of the villages have water at their feet, and that which comes from higher elevations for the irrigation of their plants, helping the chile, sweet potatoes, manioc, pacayas and other herbs...."

[3] The majority of the potatoes grown in the tablones at altitudes above Panajachel are strains introduced from Europe via North America sometime after the sixteenth century. The practice of potato raising sensu strictu seems to be a pre-Conquest survival. Bringham's (188:136) observations point to this:

"On the hillside were ancient potato fields only cultivated by digging the tubers... but it was certainly not the common potato of cultivation (S. tuberosum). The indios declared the potatoes had never been planted, but their ancestors had dug them from the remotest times, "'en todo tiempo, señor.'"

McBryde (1947:140) further corroborates this view:

In Southwest Guatemala, I found potato cultivation confined almost entirely to large "American" varieties (S. tuberosum) red and especially white, probably imported within relatively recent times. In competition with this is a small, round red potato (S. andigenum f. guatemalense), probably a pre-Columbian introduction here, little developed beyond the wild state.

Potatoes fitting the description of S. andigenus f guatemalense are available in the Panajachel market today. Perhaps indigenous Guatemalan potatoes played an important role in subsistence efforts of the highland pre-Conquest peoples, promoting use of "lazy-bed" style plots (similar to tablones) for their cultivation.

[4] Martinez (1928:22-27) offers a possible reason why farmers failed to mention the amaranths. In central Mexico these ancient seed crops were discouraged by the Spaniards "because they were bound too closely to the native religions." Also it might be inferred that the Spaniards wished to suppress the native grains in favor of European grains. Because the Spaniards were accustomed to Iberian grains, they demanded these in tribute. The same situation probably occurred in Guatemala, placing the grain amaranths on a list of proscribed or prohibited cultigens. Tobacco is the most prominent example of this legacy in Guatemala today.

[5] In reference to the chinampas of Mexico, Coe (1964:95) remarked: "The growing of flowers for sale goes back to the pre-Conquest era, when flowers were offered on the altars of pagan gods." While Coe may not have meant "sale" in the formalist "cash-nexus" or commodity sense that such products are sold today, it does raise the question whether specialized horticulture as an ecotype had advanced beyond the barter or ritual exchange basis in central Mexico. Flowers and their production for cash or for ritual obligation were probably an important stimulus for careful and intensive horticulture in Guatemala, as well.

6
The Tablón Process

Fallowing

In the usual tablón cycle, individual tablones are rested for about six months every two years. But this is not a strict practice. Many factors account for the presence of fallow tablones. Changes in ownership, lack of capital for supplies, changing employment practices among the family members, as well as strictly agronomic considerations, all play a part in the decisions that keep a tablón out of production. At the time of the survey, about 17 percent of the tablones in the transect were in temporary fallow (less than a year). Another 4 percent were "clean" (stripped of crops or weeds) and awaiting remaking before planting. Finally, an additional 1 percent were in various states of long-term fallow. These contained remnants of strawberries, sweet potatoes, beans, or other crops that had been abandoned to weed succession. Some of this latter category may escape harvest altogether, but theoretically these "abandoned" tablones could be tapped for some food in the event of extreme need. In this regard, the tablón system resembles swidden systems. Old or "abandoned" swidden plots often continue to yield produce while they are at rest. Referring to the Kekchi, Carter (1969:125) has suggested: "It is impossible to conclude when cropping ends and strict fallowing begins, for the two occur simultaneously."

Clearing

The precise clearing procedures are dependent on how long the tablones have lain idle. In the case of "new land" (though one might well question this concept in an area such as highland Guatemala where there has been continuous occupation and farming for the past several millenia) or on land that has been fallow for several decades, clearing with burning might take place. Here swidden methods could apply. This observer did not see any such land being developed for tablones while in the field. Enough land in differing states of fallow existed in the basin to preclude the need to open up forested land for tablones.

In the past, the periodic flooding of the Panajachel River has destroyed cafetales. Land deforested in this manner often becomes tablón acreage. This happened after the 1949 flood. In the transect, the land for tablones 454-715 was cleared this way. Concrete retaining walls were built on each side of the river following this flood. For the present, this means of acquiring new tablón lands seems closed. Some Indians suggested that continued deforestation of the upriver watershed will produce larger floods in the future. If this happens, it will create new lands for the tablón farmers, at the expense of the coffee growers. Tablón farmers see this as a mixed blessing. Flooding may clear cafetales, but it also creates monumental rock removal problems on some existing tablón lands (Plate 10).

Fallow tablones with brushy vegetation are cleared with machetes. The refuse is piled and burned, and the ashes are added to the soil. Fallow tablones without woody shrubs are cleared by scrapping with an azadón. This includes tablones that have been recently harvested.

Tablón Construction

After clearing, the tablones are tilled with the azadón. This operation is done by one or two men, or by father and young son. They start at one end of the tablón working up the soil in front of them for a distance of a meter or so. This aerated pile of soil is about twice as high as the cleared portion of the tablón before it. This creates a lateral trench spanning the width of the tablón. Green manure from the tablón cleaning is placed in the trench. Sometimes ground litter carried from neighboring cafetales or animal manure is added to the green manure. Standing immediately behind the trench, the process is repeated, covering the trench and working up another meter of the tablón.

After the entire length of the tablón is excavated and manured, work begins on forming the sides. With one sweeping motion, soil is scraped from the cuneta and piled on the sides of the tablón. With a second thrust, the azadón blade is used as a tamper. The sides are compactly tamped, giving the tablón the appearance of a low pyramidal base with a 60 degree slope (Plate 11). After the sides are formed and the cunetas cleared, scraped, and tamped, the tablón core is retilled (Plate 12). This fine tilling allows any remaining root crops or pebbles to be removed. As the tablón top is being retilled, it is smoothed to a horizontal plain. This is followed by still more fine combing for pebbles, bits of vegetative matter and the like. At this point the lips or "ears" of the tablón are smoothed (Plate 13). The final leveling of the top is done with a cane pole. To make a standard sized tablón (3 by 30 m) generally takes one man eight hours. This is equivalent to one tarea (a standardized task). This work is normally paid by the tarea and not by the hour.

If maize is planted on tablones, an additional task is involved. Throughout highland Guatemala, maize is mounded. Mounds, made with the azadón, are spaced slightly less than 1 m apart from apex to apex. Two widths of maize tablones occur. A 3 m tablón has three mounds; a 4 m tablón has four. Mounds are said to give maize plants stability in maturity, and some protection in their infancy from insect pests and birds.

Strawberries are another plant that require modifications of the tablón structure. Small ridges are made on top of the tablón for strawberry plants. These ridges run at uniform angles or are set perpendicularly to the long side of the tablón. The furrows between these 10 cm high ridges serve as miniature cunetas to allow irrigation water to flow in between all of the rows on top of the tablones. For this reason, strawberry tablones have low sides. Eight inches (22 cm) is normal (Plate 9). Most other tablones on sandy soils have 33-36 cm high walls. The "chinampa-tablones" have sides slightly higher than 1 m. The height of the sides is a function of crop type and soil composition.

TERRACING

Terracing is not needed in the delta as such, but the eastern slopes of the valley are extensively terraced. The terraces range in complexity from rather informal cuts in the slopes to elaborately walled and rocked structures interconnected with irrigation ditches. No new terraces were being constructed during the period of personal observation, therefore first-hand information on how much labor is needed to construct them, or whether this work is done cooperatively or individually, cannot be offered. However, each farmer is responsible for the upkeep of his own terraces and the attendant section of the irrigation system.

SPECIAL FEATURES

Natural features present occasional problems for tablón construction. Some areas within the delta have large boulders that must be left rather than removed. Sometimes these are used for güisquil arbors. In some cases, they may even function as heat conductors that moderate temperature changes between day and night (Wilken 1972:550).

Depressions are sometimes exploited rather than filled. Certain crops, especially cintula, favor damp or well-watered locations. In these cases tablones are built in the depressions. However, in places where natural springs make cultivation difficult, the land is tilled and drained. Reclaiming land in this fashion--for those willing to expend the energy--is a relatively inexpensive way to establish a "homestead."

One young man, who had lived in Sololá before migrating to the delta five years hence, had recently done this. He was initially attracted by the opportunity to sift sand and gravel by hand on the flood plain to sell to the state highway agency on a free-lance basis. After saving some money, he was able to buy two cuerdas (1,448 sq. m) of land (tablones number 272-293) which needed drainage. He built a small shack for his family on the plot.

Salvaging the land for tablón construction took four separate operations. First, all springs had to be located, and ditches dug to connect them. Second, tile--large, flat rocks--was laid. Next, gravel had to be hauled from the river to fill in the ditches. Finally, the entire surface was tilled and smoothed before tablones and irrigation ditches were built. He bought the land in 1973 for Q100.00 per cuerda. One year later he said the same land (without improvements) was worth Q150.00 per cuerda, and, with improvements, as much as Q500.00 per cuerda. He felt lucky to acquire land so cheaply; land that now has a manageable and permanent water source in situ.

PLANTING

Seed Planting

A majority of the tablón crops are planted from seed. Some are planted vegetatively from cuttings, while the rest are self-propagated and adventive. Seeds for European-introduced vegetables like beets, cabbage, lettuce, carrots, and turnips are imported. Many of the other seed-propagated crops are allowed to "go to seed." The seeds are then collected. For example, Panajachel is a center of onion seed production. Panajachel's onion seeds are sold to other communities around the lake, and to other horticultural centers outside the basin. Each crop has different propagating procedures involving spacing, timing, and so forth.

Onions--as well as beets, carrots, cabbage, and others--are seeded first in beds, or semilleros. Generally, these beds are located at one end of the tablón. In addition, as previously described, some rock piles are used as planters, or, in a few cases, tablones "upon tablones" are constructed with the seed bed occupying the upper tier. The seeds are broadcast after preparatory watering and soil manicure. A 10 mm film of damp soil is sprinkled over the seeds, and the seed bed is covered with pine needles. Sometimes ilamo leaves or banana fronds are used as coverings. The covering is watered, and, if it is an onion seed bed, lightly watered daily for nine or ten days. After this, the covering is removed; the starts are tended until ready for transplanting.

Transplanting usually takes place after two months. Women kneeling on pallets (22 by 40 cm) to minimize damage (by spreading their weight over an increased area) dig up the seedlings with palitos. The roots are trimmed, and the seedlings replanted. Tax (1953:53) gives a distance of 10 cm between transplanted onions. Today, it is more common to space onions 5 cm apart on the first transplanting, 10 cm on the second, and 13 cm if there is to be a third relocation. Closer spacing and multiple transplanting is routine today. This suggests that an intensification of methods has taken place during the last few decades.

Techniques for transplanting other vegetables started in beds varies according to the plant. All transplantings are given individual attention as horticulture demands, though none as intensively as onions. Vegetables grown from seed are planted throughout the year. To this extent, they serve an interstitial role in the main seasonal "dialectic" between the "winter" planting of maize and the "summer" plantings of garlic and beans.

Garlic and beans are planted at roughly the same time--after the maize is harvested. While beans are planted from seed, garlic is planted from cloves. Garlic is spaced about 10 cm apart. Beans are spaced "un mano," or one hand apart, which equals the distance between the outstretched forefinger and thumb--about 20 cm. Watering during the dry season ("summer") beginning in November is done every three or four days. Both crops take about 6 months to mature.

Tuber Planting

Manioc, sweet potatoes, and arracacha are all propagated vegetatively from cuttings. Manioc is grown from sections of the stalk placed on the tablón tops in rows every four or more feet. At a distance they can be mistaken for the sharpened stakes sometimes placed at 40 degree angles around the perimeter of tablones to discourage roaming dogs. Sweet potatoes are planted from cuttings placed in the tablón sides every 25-50 cm. Arracacha is grown in a similar manner. Tubers, which may be planted any time during the year, are in continuous production supplying the farmer with a dependable starch food at any time after they approach maturity.

Other Crops

The remaining 100 or more tablón plants fall into two planting categories. Many of the herbs and greens are volunteers which are simply selected and tended during the weeding process. Many of the tree crops are adventive as well. On the other hand, some of the herbs and greens that may be adventive in their initial appearance are allowed to go to seed, and their seed is collected for replanting and consumption. Hierba blanca, a common native mustard, is allowed to "go to seed" in this manner. It occurs in 24 percent of all tablones. The seeds are used as a condiment, the leaves are eaten as a green and a condiment, and the seeds are broadcast for reseeding in tablones. A number of semidomesticated herbs and greens are exploited in a similar manner. No particular spacing arrangements are followed with these semidomesticates, however European herbs and spices, like borrage (Borago officialis), or the various cultivated flowers are usually grown along the sides of tablones. In general, spices, herbs, and flowers are planted randomly or where space allows.

WEEDING

Weeding is a labor-intensive task done about twice a month for most crops. It is either done by women and children, or the poorest stratum of the region's landless laborers. These laborers are paid Q0.25 per day for women, Q0.50 for men. Indians from villages above the lake from as far away as 6 km walk daily to the tablones to solicit work weeding. Weeding, like transplanting, generally requires kneeling on top of the tablones. Small boards are used to distribute the weeder's weight evenly. The weeding is done with the palito. In general, tablón farmers keep their plots well-weeded, even though close planting and alinear arrangement makes weeding more difficult.

FERTILIZING

Care of maturing crops is usually carried out along with weeding tasks. Little, if any, mulching is practiced. Presumably, space economic considerations prevent mulching--most tablón crops are spaced as tightly as possible. They grow on beds--not in rows. Therefore, there is little additional space for mulching between the plants. Mulching is obviated by frequent weedings. Thinning and pruning is usually done while crops are being transplanted. Since green manuring is carried out when the tablón "is made," the only application of fertilizer involves chemical means while plants are maturing. The most common method of applying chemical fertilizer requires perforation of the tablón surface with a palito. Granules of fertilizer are dropped into the perforations. This "gives force" to the crop, if it is thought to be in need of help, or if the farmer has extra fertilizer on hand.

Experimentation with chemical fertilizers is very much in its infancy in Panajachel. Many farmers have not learned the delicate nuances of chemical fertilization in a system that approaches the pedologic simplicity of hydroponic gardening. One frequently heard complaint is that chemical fertilizers have so changed the soil that "the land is now addicted to chemical fertilizers in the way some naturales, "Indians," become habituated to alcohol." Generally only one solution is seen: apply greater amounts of fertilizer, even though this may be counterproductive. To stop chemical fertilization is to invite unacceptably low yields. The farmers are painfully aware that they have temporarily altered the soil compositions in many of their tablones beyond a point of immediate restoration. Few farmers have the capital to allow their tablones to lapse into fallow or to restore them with organic mulches and manure. This dilemma will probably drive more and more farmers to intensify production through monocropping of onions and other cash crops that are highly space economic. Of course, the circle may "close" as a result of increased production: the price of onions will drop. The Indians see the Ladino vender of chemical fertilizer and the Ladino buyer of onions as the "winners." This does little more than confirm the Indians' conception of their position vis à vis Ladinos for the past four centuries.

YIELDS AND RETURNS

Returns from tablón culture vary according to a number of factors. Therefore it is difficult to state with certainty the returns for a single individual, let alone the community as a whole. Nevertheless, Tax (1953) made a rather extraordinary attempt to do just that. It cannot be pretended that the data gathered for this present study allows the author to make no more than a rough estimate of the returns from one crop: onions. The following estimate is based on costs and yield maxima extrapolated from information given by several farmers. The accuracy of their information may even be questioned, because in some cases this student may have been considered more than a disinterested observer--and quite possibly a potential, if inexperienced, buyer. Every effort has been made to take this into consideration. Table 6.1, at the end of this chatper, outlines the labor requirements for growing one standard-sized tablón of onions.

Our idealized example involves a farmer who rents his land. Rent per cuerda per year may be as high as Q12.00 for sandy delta land. Using a standard 3 by 30 m tablón as our unit, rent for one tablón for one crop (four months) would be Q0.41.

Assuming the farmer hires a "mozo" to make the tablón, the cost would be Q1.25. Seed costs as much as Q7.00 per pound. To plant a standard tablón from seed grown in a seed bed (assuming the farmer does this himself) would take 1/5 pound, or Q1.40 worth of seed. Wood ash applied to discourage grubs, especially the "gallina ciega" (Melolontha spp.), costs Q0.25 per bag. Three bags would probably be used, for an additional Q0.50. Later applications of insecticide might be used, costing about Q0.75 per tablón. We will assume only one spraying, though more is often done. Chemical fertilizer sold for Q0.18 per pound at the time of the study. Five pounds in each of three applications was one recipe. While not a conservative amount to be applied to one crop of onions, this amount is exceeded by some farmers. The cost would be Q2.70. If the farmer hired outside help to do the weeding, two weedings would cost Q2.00. This is based on two women working two days to weed one tablón, at Q0.25 per day per woman.

If the farmer followed traditional practice he might hire a shaman, or zahorín, to say a costumbre for his plantings. This service might cover his milpa as well. Retainer for the zahorín, depending on the shaman's own reputation and on the elaborateness of the ceremony, might be as high as Q20.00. Supplies for the costumbre (chocolate, beer, aguardiente, tobacco, incense, candles, flowers, etc.) might run an additional Q20.00. Given that the farmer has one cuerda of tablones under cultivation, and one acre of milpa, the "cost" of the ritual for the single tablón would be about Q0.70. Therefore, total expense--excluding the cost of the labor of the farmer and his family involved in tasks like irrigation, guarding, etc.-- would be Q10.21.

This idealized tablón of 9,000 small onions (25 to 30 mm in diameter), planted and harvested to maximize space economy, might sell for Q20.00. If the farmer decided to harvest, wash, sort, and bundle the onions, then the price could be as high as Q30.00. Clearly, the profit--from Q10.00 to Q20.00--is not much if the farmer's labor is included. In rendering this estimate, we have tried to maximize all figures. Even in the best of times, the costs do not decrease much. In the worst of times, the sale price may plummet to half or less than the prices given above. Discontent among tablón farmers over this situation is not surprising. In most cases, the alternative--coastal labor--is even less appealing.

Because this investigator was in the field for only the fall season, it was impossible to get complete information on yields and returns from other crops. The reader is referred to Tax (1953) for nearly complete data on yields and returns in 1936. One can only reiterate the impression that root crops and tablón beans, as well as many seemingly insignificant herbs and greens, play a central role in the tablón farmers subsistence efforts. For this assumption the farmers must be taken at their own word.

HARVESTING

There are various strategies for harvesting tablón crops, largely depending on whether they are bound for market, or destined to be consumed by the farmer and his family. Onions and garlic are often sold by the tablón to middlemen who do the harvesting and bundling, if not actual marketing. Strawberries are most often picked by the women, and sold directly on the street by the young women of the family. Corn is the only crop in which the communal harvesting strategy familiar in swidden systems takes place. Even this rarely involves persons beyond the extended family unit. Other crops are harvested much less formally. Some are picked when needed. Others, like root crops, are collected as the tablones are being remade for planting. Paying outsiders to do the harvesting does not seem to be as common as hiring help to do weeding or tablón making.

MARKETING

Tracing the dispersion of Panajachel's tablón produce beyond the town raises a geographical question of some interest. Information gathered on this question suggests that there is a correlation between the distance from Panajachel a crop is marketed, and the amount of land devoted to its cultivation. For example, Sapper (1897:305) encountered Ixil traders marketing garlic and onions as far away as Tabasco. Encountering the same thing, Thompson (1964:22) thought that this produce came from the Sololá/ Panajachel hearth, a distance of some 400 km. McBryde (1934:124) saw onions from Panajachel in transit to Tapachula on the Mexican border, for later trans-shipment into Mexico. Many of the older Panajacheleños spoke of making obligatory trading trips to Guatemala City in their youth. This amounted to a coercive corvée system. The trip took four or five days by foot. The main items carried were onions, garlic, pepinos, or avocados in 45 to 70 kg lots.

Currently, Panajachel's onions and garlic, as well as lesser crops, are traded in large amounts throughout southwestern Guatemala and in the neighboring countries. From available records, it is difficult to date the antiquity of this trade. However, if ancient Mayan trading patterns for other commodities are representative (Thompson 1964), one might postulate pre-Columbian trade in horticultural produce from this area. In any event, onions and garlic are not only the most predominant tablón crops, they are probably the most far ranging crops marketed. This question deserves more study.

Other tablón crops enter the market in lesser amounts, though their contribution to the town's economy is significant. Tree crops from Panajachel are marketed widely. Panajachel's avocados are harvested and marketed by Pableños. The men from San Pablo combine avocado picking and marketing with their cordage-making specialization. Contracting with individual tree owners, they periodically harvest the town's avocado crop and take it to Guatemala City. Similarly, Panajacheleños and free-lancers from other villages do the same with a variety of fruit-bearing trees in Panajachel. At one time, Panajachel was the center of anise culture in Guatemala. Today, San Antonio has that distinction, but when Panajachel was the center of this crop, it was marketed great distances. Perhaps dozens of other crops could be similarly cited. The main point is that Panajachel produces many crops in excess of its own local needs. These are marketed directly from Panajachel to villages throughout Guatemala and beyond, or they are taken to Sololá where they enter the regional marketing system on a slightly more centralized basis.

Table 6.1
Labor Requirements for Growing One Standard-sized Tablón of Onions*

TASKS	TIME SPENT	TOTAL HOURS
A. PREPARATION		
Cleaning and clearing	1.0	
Scraping (with azadón blade)	0.5	
Tablón construction	8.0	
Pre-seeding soil manicuring/watering	1.0	10.5
B. PLANTING		
Seeding**	3.0	
Spreading top soil covering	0.5	
Mulching (for seedling protection--using pine needles)	1.0	4.5
C. TRANSPLANTING		
Digging up seedlings	8.0	
Trimming roots	2.0	
First transplanting	8.0	
Trimming roots	2.0	
Digging up onions for second transplant	8.0	
Seconding transplanting	8.0	36.0
D. FERTILIZING/WEEDING		
Chemical fertilizing (3 applications)	12.0	
Weeding (12 hour sessions)	36.0	48.0
E. WATERING		
Seedlings (generally watered for 9 consecutive days)	18.0	
Onions nine days after germination***	80.0	98.0
F. HARVESTING/PREPARATION FOR MARKET		
Harvesting	8.0	
Washing	3.0	
Hauling to staging area for sorting/bundling	1.0	
Sorting and bundling yield of one tablón	4.5	16.5
TOTAL NUMBER OF HOURS		213.5

*Standard-sized tablón=90 square meters; growing time=129 days.
**Onion seeds are sown with more care than simple broadcasting; seeds are thrown
  individually onto the graded tablón surface.
***This assumes a four-month growing time.

7
Irrigation

THEORETICAL ASPECTS

Three classes of construction figure prominently in the tablón landscape. The first two, outbuildings and fencing, are more than embellishments, but less than essential constructions in the tablón system. On the other hand, irrigation works are essential to the system. Tablones, as they exist in Panajachel and elsewhere in Guatemala, are adaptations to seasonal aridity. The construction and maintenance of the tablón irrigation system is largely a communal affair, in contradistinction to most other tablón-associated tasks. Much has been written on the role of irrigation as a centralizing force in agricultural societies (Wittfogel 1957; Steward et al. 1955; Downing and Gibson 1975). Given the scale tablones occur within the Lake Atitlán basin, one might ask the question: How intrinsic is the link between societal centralization and irrigation? The opposite argument is more convincing: That decentralization of power and mutual aid are characteristic expressions of small-scale irrigation systems, and perhaps even allow these communities to resist centralizing authority (Butzer 1976).[1]

WATER SOURCES

The irrigation system is fed from two sources. The Panajachel River is bled near the north end of town for the west side. The Tzalá River and several springs are tapped for the east side of town. Tablones in the southwest corner of Jucanyá are also irrigated from the Panajachel River. The entire system relies on gravity, except for the tablones adjacent to the chinampas in the center of town. Often these latter tablones must be irrigated by pushing water into their tomas from the water standing in the chinampa ditches. The azadón blade is used like a piston in the toma walls when water flowage from the chinampa springs is inadequate (Plate 15).

CANAL TYPOLOGY

Water bled from springs or the two rivers travels through a network of ditches to the fields. These ditches are called tomas grandes, and usually follow property lines. They are the largest irrigation ditches in a five-part ordination. Water flows in the tomas grandes all year. Diverting the flow completely from one of the tomas grandes would be considered an anti-social act. It would not be likely to occur. The tomas secundarias, on the other hand, carry water only part of the time, directing water to a given section of tablones as needed. Tomas de tablones are ranked third in terms of the volume of water carried. These tomas divide two rows of tablones. The toma de tablón may service from 2 to 40 or 50 tablones, though the average number of tablones in one block is around 20. Cunetas, the lateral ditches between two tablones, are remade each time the tablón is formed. Finally, in the case of strawberries, cunetillas are formed by the strawberry rows on top of the tablones.

The size of the tomas grandes vary according to the local relief, i.e., whether the gradient is level or steep. The largest tomas are 1.5 m wide by 1 m deep and resemble small brooks. The tomas secundarias are about one half the size of the free-flowing major tomas. Tomas de tablones and cunetas usually have standard dimensions. The width (33 cm) at the bottom corresponds to the width of an azadón and the distance between the two tablón ears at the top. This distance varies according to the height of the tablón sides and the slope of the sides. Tablones in sandy soil with 27 cm high sides would probably have a cuneta of 30-33 cm at the bottom, and 60-65 cm from tablón ear to tablón ear.

The smaller links in the system--the tomas de tablones and the cunetas--are rarely filled to capacity. Only enough water to comfortably splash irrigate (15-20 cm) is normally passed through the system. On the other hand, when the cunetillas (10 cm deep--the height of the strawberry rows--and 15 cm wide) are used, they are filled to overflowing. This means that the cunetas also must be filled to capacity. Therefore, strawberries are particularly demanding in the sense that they require more water to be flushed through the system than other plants.

One ingenious adaptation falls outside of this ordination. In the areas where terraced tablones are built on steep slopes, fitted banana frond conduits supported by cane poles are used to transport water across small saddles and around boulders (Plate 14). Using this portable system, water from small cataracts is channeled to narrow tablones. As previously mentioned, most of these tablones are planted in pole beans to maximize the use of horizontal space.

The labor to build the irrigation system is communal in the case of the two largest orders of ditches. Labor to build the three smaller orders of ditches is organized and executed on an individual basis. This also applies for the banana frond conduits. Maintenance for the larger tomas is an annual affair. Theoretically, all resident agriculturalists participate. Exemption can be purchased by providing a day laborer to take one's place, but the festive spirit encourages most agriculturalists to join in this day of village-wide communal labor. Gangs of six to seven men travel the entire network, cleaning and repairing the ditches. Failure to participate results in a fine levied by the municipality. By custom, the day set aside for this work is a Sunday at the beginning of the dry season, usually at the beginning of November. However, in November 1974, Thursday the 21st was chosen. Some 70-80 men participated. Perhaps the choice of a weekday had some effect on the turnout, but the task was completed in one day, even without all available agriculturalists working on it.

In some instances, secondary tomas are built to provide new sections of tablones built on land not previously farmed. In this instance, the farmer developing the tablón plat will enlist the aid of his compadres, or the male members of his extended family, to build the secondary ditch. Access across the intervening land is not paid for, but permission is asked of the owner(s). Apparently, water is plentiful enough to allow for informal enlargements of the system without consultation with the farming community as a whole. This practice leads to the speculation that the manner in which the system was originally constructed was by ad hoc decision-making involving groups of individuals, rather than by bureaucratic design.

IRRIGATION ACTIVITIES

Watering the tablones from November through May, and during the canícula, greatly extends the productivity of the system. Horticultural efforts in neighboring villages are less than half as productive if they rely on rainfall alone. For the villages that practice it, irrigation is a bi- or tri-weekly task during the arid months. Watering is done almost exclusively during the late afternoon, night, or early morning. Watering during the heat of the day risks damaging the crops, and is not an efficient use of water resources because of evaporation. Except in a few cases—moistening a tablón before remaking, or watering strawberry tablones—tablones are not flooded, but splash irrigated. Splash irrigation is done with palanganas, usually by women and children. Standing in the toma and bending over, the irrigator scoops water from the ditch and splashes it onto the tablón. A fine spray is assured in this way. Soil surface features are only minimally disturbed as a consequence.

The splashing is done in rhythmic flurries lasting 20-30 seconds (at the rate of 70-80 palanganas per minute). The volume of water spread in this manner averages about 35 liters per minute. Every few minutes the irrigator moves a few feet along the ditch to water a different spot. For almost all crops, watering is done every three days during the dry season. A standard sized onion tablón receives about 4,200 liters of water per week. The work involved in watering tablones markedly increases the amount of labor committed to tablón culture as a system. Almost half of the total labor time expended involves irrigation.

NOTES

[1] The development of specialized horticulture in the shadow of larger urban centers within Guatemala seems to function as an avoidance mechanism. Through intensive cash cropping (vegetables for the national market), tablón farmers are generating in miniature--while at the same time trying to escape--the pressures of commodity cropping for the international market.

Wittfogel's model (in the case of Panajachel) may fit the produce more than the producers. A "hydraulic organization" of space and society is evident--but it applies to the vegetables--not the peasants. If there is a "totalism" to be found in tablón culture, it is the drift toward monoculture and closer "packing" of individual crop plants upon the tablones. Geertz (1963) described the "involution" of the sawah system in similar terms.

8
Other Economic Pursuits

GUARDING TABLONES

Guarding the tablones from predators takes some time, even though it does not seem to be a major concern. This is in contrast to the situation of shifting cultivators in lowland Guatemala (Carter 1969). Farmers rarely spend the night in the tablones for the expressed purpose of discouraging pests. Because they spend much of their day in the tablones, and most tablones are close to human habitations, troublesome animals are discouraged from frequenting tablón areas.

The tepesquinte, or paca (Cuniculus paca), and the pocket gopher (Macro-geomys spp.) are said to be nuisances, damaging vegetables and corn. The tepesquinte, in turn, is highly sought after as food. Its numbers are limited by human predation. Various blackbirds and crows pose problems and are generally discouraged by boys armed with small slingshots, not very similar to the standard Euro-American slingshot. They are effective hunters and keep damage done by avian pests to a minimum.

HUNTING

Both hunting and fishing are more important to the subsistence effort of Panajacheleños than has been commonly thought (McBryde 1947; Tax 1953). Recent studies of "primitive" agriculturalists in Latin America suggest that the importance of hunting and fishing foci (Bennett 1962; Denevan 1966, 1974; Nietschmann 1971) has been underestimated. Similarly, most Latin American peasant farmers rely on wildlife for part of their protein needs. Even the intensive agriculturalists in heavily populated districts, like Xochimilco or Tlaxcala, Mexico, draw on aquatic animals for protein supplements to their diets (Wilken 1970).

The habitat around Panajachel supports a diverse fauna. Deer are the main large game animal. They are sometimes trapped, but not generally hunted, because Indian ownership of guns is restricted.[1] A variety of smaller game animals are actively sought by adolescents and younger boys. They use slingshots or snares to kill rabbits, squirrels, tepesquinte, armadillos, and opossums. Using the same methods, a number of birds are killed. Predictably, "game" birds like quail, pigeons, and doves are especially prized, but song birds, jays, and parakeets are killed and eaten. Several boys make a business of killing passerines (song birds) for their salable plumage and small protein returns.

All waterfowl are considered edible. Before recent conservation efforts lessened the practice, the endangered Atitlán grebe was hunted for meat. Passing ducks and other waterfowl are given no quarter if they can be caught. L. C. Stuart (1974) said that some reptiles and amphibians are eaten by Panajacheleños.

Much of the hunting is spontaneous, done as the occasion presents itself while performing other tasks: farming, fishing, traveling. Most of the other hunting occurs in an avocational context. The few examples of young boys devoting the majority of their non-recreational time to hunting involved ladinoized Indians or Ladinos. They specialized in trapping wild squabs in the mountains for cage birds or for food. One group of Ladino boys claimed to have captured 100 pigeons per month by stunning them with their slingshots from blinds built in trees. They allegedly sold the birds for Q1.00 each. If this was accurate, they netted their families a respectable amount of additional income. While the stated price paid seems inflated, the numbers mentioned are plausible. In sum, Panajacheleños are not solely agriculturalists. Hunting remains a way of supplementing their protein consumption.

FISHING

Another faulty generalization is the notion that Panajacheleños do not fish, leaving this activity to other villages.[2] While this may have been true in the past, today many Panajacheleños fish on occasion. This activity does not assume the same level of specialization as it does in other villages (most notably Santa Catarina), but, to characterize Panajachel as a village without a fishing focus is inaccurate (Plate 3).

At least one family combines tablón farming and spear fishing. The use of skin-diving equipment to hunt large-mouth bass has become the hallmark of this family. The elder sons learned the practice from Ladino vacationers.

Bass meat is sold to local hotels for Q1.50 per kg. Bass ranging from 2 to 6 kg are caught, with even larger bass having been reported. The equivalent of one week's wage in agricultural work can be earned in a lucky encounter with a large bass. It may be that this particular family, by specializing in bass fishing for the hotel trade, can provide the capital for equipment and allow the elder sons to fish because they are successful agriculturalists. Several families fish with hook and line for sunfish and bass. Fish are sold to the hotels or marketed from door to door. Many younger boys and some older men spend a portion of the early morning and late afternoon fishing from the shore for small bass and blue-gill. These fish are consumed directly by the anglers and their families.

Open-lake fishing by canoe is the predominant mode practiced by the villages specializing in fishing and crabbing. Some Panajacheleños own canoes, but the fleet is much smaller than Santa Catarina's. There is some evidence that fishing in general is becoming less profitable. This is because the larger bass have been depleted. In turn, bass and other exotics were responsible for the collapse of the native fishery.[3] One might speculate that if this trend continues, the villagers in Santa Catarina may be forced to expand their meager tablón plots, and send more of their numbers into the streets of neighboring villages to beg from tourists. To the relatively staid agriculturalist of Panajachel, the Catariñecos are the gypsies of the Basin, not to be entirely trusted, nor to be emulated. This may account for the little interest Panajacheleños have shown for developing a canoe fishery. Maintaining local identity through occupational specialization is still a force determining vocational choices, though this is changing in Panajachel, as Hinshaw has recently shown (1975).

GATHERING ACTIVITIES

Plant Collecting

Gathering activities are varied and difficult to monitor. There are two main categories: vegetal products collected for food, and wood collected for fuel. A number of plants are collected for their medicinal qualities. Wild herbs and greens are collected to supplement those grown on tablones, or in some cases, collected in the wild because they are not grown on tablones. Edible mushrooms are collected in the mountains after a rain. Certain mushrooms are known to have distinct seasons when they are most plentiful. During these seasons, several species of mushrooms appear in the market. Wild maguey (Agave spp.), another plant found in the common lands above the village, is used for several purposes: cordage, juice, food. Wild fruits and berries are among the other foraged foods that come from the "monte."

Firewood

Though foraged food does not form the basis of the subsistence economy, foraged firewood provides the bulk of the town's fuel needs. Firewood is constantly collected from the hill slopes and from the river bank. During the rainy season, much of the firewood is brought down from the mountains by the river.

In the wake of the flooding caused by Hurricane Fifi, this author surveyed the amount of firewood salvaged by Panajacheleños. Even while the river was at peak flood stage, dozens of townspeople were out collecting wood and cutting it into transportable sections. During a two-hour period on the second day of the flooding, 10 people collected 13 cargas (6.5 cubic m) of wood from the riverbank. This observation was taken along one of the dozen major east/west trails that cross the river. The amount collected was much more for the first day. Given that on both days the collecting was done from dawn to dusk, a conservative estimate of the amount collected would be about 1,200 cargas. Cargas sell for Q0.75-Q1.00. Probably more than Q1,000.00 worth of firewood was collected during a two-day period. Used for cooking purposes only, a carga of wood may last a family one or two weeks. This particular flood brought considerable returns without much damage to most Panajacheleños.

Firewood not taken from the river, or collected as driftwood from the lake, comes from the mountains. If an Indian is wealthy enough, he buys wood by the tarea from commercial vendors who sell from trucks. A tarea is about 20 cargas, but sells for Q5.00-Q7.00, representing a savings for the Indian able to spend that amount at one time. More commonly, the tablón farmer buys wood by the carga from wood cutters who bring wood to town on a daily basis, going from door to door to sell it. Wood cutters may travel 10-20 km on foot in the course of their daily work, but usually earn less than Q1.00 for their effort. Again, this is an example of tablón farmers paying for a service that they themselves might do, but prefer to patronize a wood higgler. Still more common is the incessant scavenging of twigs and branches and other fuel on a daily basis by women and children. Potential fuel is seen in terms of quetzales and centavos; it is by no means ignored.

NOTES

[1] Termer (1957:92) described the "antiquated method" still used to trap deer. Deer are driven from the uplands down barrancas with outlets into the lake. Once forced into the water they are taken from waiting canoes.

[2] Tax's (1953:29) assessment is representative.

[3] Apparently overfishing is not a new problem. In the seventeenth century Diego de Ocana (1933:297) reported on the state of the lake fishery:

...the fishery gives crabs and minnows in abundance. The minnows are the size of the little finger and others are less than half that size. They provide the bulk of the sustenance of the neighboring villages [Santa Catarina?] and are its major source of income. The larger fish are smoked, and strung on thick straws (like barley) and put on strong sticks, and they are carried and sold in many places and provinces...

He goes on to speak of fish four and one quarter fingers long (20-25 cm) but they were "getting more scarce with the passage of time."

In the 1820s, Juarros (1823:91) did not mention the larger fish in his report on the lake region:

...the only fish caught in it are crabs, and a species of small fish about the size of the little finger; these are in such countless myriads, that the inhabitants of all the ten surrounding villages carry on a considerable fishery for them...

9
Conclusions

One of the major assertions of world social science is that there are some great
watersheds in the history of man. One such generally recognized watershed, though
one... studied by only a minority of social scientists, is the so-called neolithic
or agricultural revolution. The other great watershed is the creation of the modern world.... The mark of the modern world is the imagination of its profiteers
and the counter-assertiveness of the oppressed. Exploitation and the refusal to
accept exploitation as either inevitable or just constitute the continuing
antimony of the modern era, joined together in a dialectic which has far from
reached its climax in the twentieth century.

Immanuel Wallerstein, 1974
The Modern World System

TABLON CULTURE: A SYNTHESIS OF PEASANT ECOTYPES

This study examines aspects of tablón culture, which is primarily a fusion of two ecotypes: specialized horticulture and New
World hydraulic agriculture. Specialized horticulture is defined by
Wolf (1966:33) as "the production of garden crops, tree crops, or
vineyard crops, in permanently maintained plots." This ecotype
originated in the Mediterranean area, fostered by the "tendency
towards regional specialization along the shores of a sea linked by
maritime traffic... with historic continuity from 1000 B.C." Similar
attributes are found in varying degrees in the tablón-associated
horticulture of Panajachel. Microregional specialization is a prominent feature of tablón culture in the Atitlán area. Each village
has a particular specialization, and the villages are connected by
maritime traffic around the lake. The degree of historical continuity of tablones is debatable; though it may surpass a millenium.

Wolf's (1966:19-37) typology of peasant ecotypes is helpful in
placing tablón culture within a broader comparative context.[1] After specialized horticulture, the tablón system most clearly
evinces aspects of New World hydraulic systems. A variation of
the pre-Conquest hydraulic agriculture probably served as the
base of the present system. During the past century, coffee

cultivation, a conspicuous example of tropical commodity crop production, has been partially integrated into the tablón system. In addition, tablón culture is spatially contiguous with and/or culturally integrated into the area's long fallow (swidden) systems, sectorial fallowing (barbecho) systems, and small-scaled dairying operations. The tablón system combines six of Wolf's nine major peasant ecotypes.

McBryde (1947:3) remarked: "The highest degree of microregional diversification anywhere in Guatemala is to be found here; it is probably not exceeded elsewhere in the World." The existence of six peasant ecotypes in one valley, or the integration of five of these ecotypes within a 200 hectare area or less, lends support to McBryde's assertion. However, in geography conceived as social science, it is no longer enough to produce monographs of peoples and places with descriptions of cultural diversity as the sole objective. Brookfield (1973:10) has suggested that "all attempts to establish a valid intermediate standpoint between the unique and the universal demand that we study the nature of variations in a cautiously comparative manner."

BEYOND "INVOLUTION"

Geertz (1963) argued that the two most prominent Indonesian ecotypes--swidden (shifting cultivation) and sawah (permanent wet rice cultivation)--could be used as metaphoric lenses to view that country's agricultural development under colonialism. In part, the same method can be applied to Guatemala. Rather than single out two distinct ecotypes and focus on their separate trajectories as Geertz has done, the Guatemalan situation requires a more synthetic view. This does not mean that Guatemala is less bifurcated into an Inner and Outer realm than the division Geertz described for Indonesia. Simply put, there is no agricultural mode of production in Guatemala that entirely corresponds to the "involutionary" character of the sawah system.[2] Therefore, we must look elsewhere in Guatemala for the "plus ça change" principle of colonial agriculture.

Of the peasant ecotypes in Guatemala, tablón horticulture most closely resembles the involutionary nature of the sawah system. The major difference is that unlike sawah farming, tablones do not support ever greater populations. Because tablón culture is an integration of several ecotypes, there are competing and contradictory demands placed on the system. Tablón culture is only involutionary on the level of the individual tablón.[3] At this level, the pressures of spatial circumscription of land on the delta, and market demands, encourage farmers to place individual plants as closely together as possible. Plants that allow maximum vertical "packing" are "selected for" in the system. Onions and garlic maximize space-efficiency in Guatemalan tablones. Wet rice represents their analogue in Indonesian sawahs.

At the second level of the hierarchy--the local level--there is a drift toward monoculture in the southwest sector of the delta. In this sector onions and garlic are more nearly monocropped than elsewhere in Panajachel. In part, this constitutes the beginnings of an involutionary pattern, but this pattern has yet to "crystallize" on the local level as a whole. The local level demonstrates a fairly stable interplay among competing ecotypes over time. For example, coffee acreage (representing the tropical commodity-crop ecotype) has not changed much since the 1930s; nor has the amount of land devoted to specialized horticulture. Milpa (swidden) acreage has declined somewhat (Hinshaw 1968:72).

At the regional level, specialized horticulture is the dominant and most dynamic ecotype. Its expansion since 1900 has been a dual process. First, tablones have expanded to new locations on the lake, and they have expanded vertically to include more of the basin's steep slopes. At the same time there has been a tendency toward monocultural specialization. Increasingly, villages that once specialized in a variety of crops are turning to a few market items--particularly onions and garlic.

The interaction between ecotypes at the national level is well defined. There is a marked division between Inner and Outer Guatemala. Inner Guatemala is typified by the Indians and their dominant ecotypes: sectorial fallowing and tablón culture. Outer Guatemala is the Ladino realm characterized by swidden, some dairying, and extensive plantation cropping of commodities for the world market.

The national configuration (combining the "two Guatemalas") has been determined by forces beyond the national level. Guatemala's role in the global division of labor is quite similar to many other tropical countries. The predominate ecotype is production of plantation crops for the world market.[4]

Evidence of the interplay among ecotypes at all five levels of articulation can be seen at the two most basic levels: the individual tablón and the local level. For this reason, this author concentrated on the formation and transformation of tablón culture in one location. This approach has generated an hypothesis of change within the tablón system of Panajachel.[5]

This hypothesis, or "explanation sketch," may apply to many situations wherein specialized horticulture is implanted upon an older base involving hydraulic agriculture. This seems to be a reasonable explanation of the formation of tablón culture in Panajachel. Presumably, this process has occurred elsewhere in Mesoamerica and in other parts of the world. As Brookfield (1972:44) suggests, after micro-geographical studies of individual systems, we should turn to pan-tropical comparisons. Tablón culture represents a system of some antiquity and demonstrates a fusion of several peasant ecotypes. Cognate systems should be identified and comparatively studied.

SUMMARY OBSERVATIONS

Four main questions have been examined in this study:

(1) What are the functional characteristics of the tablón system?

(2) What are the origins of this system?

(3) What role does tablón culture play in promoting economic specialization within the Atitlán Basin, and in blurring the boundaries between Inner and Outer Guatemala?

(4) What directions might future theoretical work on the nature of the tablón system take?

<u>Functions</u>

The first question has been approached with description of selected features of the system. This descriptive sketch is incomplete. Many important aspects of the system were ignored. For example, physical processes involving soil, hydraulic circulations, microclimatic features, pest problems, and so on remain undescribed. Instead, a thorough attempt was made to record crop composition. In a limited and synchronic sense this was successful. A diachronic analysis might have involved one or more complete field seasons and a more complete study of the historical materials. Some attempt has been made to examine crop changes over time; especially as these have occurred at the regional level. The rationale for focusing on crops almost exclusively (beyond the expediencies of time and expertise) is linked to what this author believes Steward (1955) meant in arguing for the study of "cultural cores," or those activities most directly related to subsistence activities. Crop compositions (and the processes that regulate them) are at the core of the tablón culture.

Veblen (1974) correctly recognized the centrality of "crop focus" in his argument that tablones are post-Conquest artifacts. While this present study suggests an equally good case can be made for a pre-Conquest origin of tablones, the question of crop composition remains central. There is enough recorded information to examine changes in crop compositions over time and the diffusion of cropping preferences within the region. My data suggests that increased specialization in cropping has occurred. This specialization emphasizes "space-economic" plants like onions and garlic. The trend toward space efficiency in farming is similar to settlement patterns in the region. Level land within the basin is scarce, therefore nucleation of settlement and subsistence patterns is extreme.

Harris (1974) has argued that one need not make a fetish out of the complexities of cultural processes. These "complexities" often result more from demands originating in "scholarly subsistence systems," than from the realities of the subsistence systems being studied. The system must seem all too simple to the farmers of Panajachel. As Smith (1972:234) observed: "Land pressure has only made subsistence farming rather than all farming unviable." In the core areas of Inner Guatemala (like the Atitlán basin) the market system has forced the peasants to "turn to horticulture or skirt-cloth production." Land is scarce and increasingly expensive; prices on goods near to the Ladino market centers are high. For the farmer, the solution seems equally uncomplicated: he must intensify horticultural production energetically, spatially, and in terms of crop composition.

Origins

If the pressures pushing the tablón system toward greater specialization and intensification seem straight-forward, the agencies that created tablón culture are more obscure. The confusion and inability to see tablones as pre-Conquest artifacts arises from the implied assumption that the tools, plants, and techniques were largely imported from Spain. This could be accurate. It seems more reasonable to assume that tablón culture is a fusion of several peasant ecotypes. Elements of Mediterranean specialized horticulture are prominent. These elements are laid upon a more ancient base, namely New World permanent or hydraulic cultivation. Tablón culture is complicated by the presence of other ecotypes in proximity: swidden, sectorial fallowing, dairying, and tropical commodity crop production. This last ecotype has a quasi-adversarial relationship to the tablón system in Panajachel (Tax 1953). In sum, the Ladino impress of much of current tablón culture has obscured its probable pre-Conquest origins.

New Theoretical Directions

This raises the final question: How much has current theory obscured the role of the residua of "pre-capitalist modes of production" in the formation of current peasant ecotypes? This problem is one of Marx's most famous lacunae. Since Marx, the question of "Asiatic" or "Oriental" modes of production often has been associated with Wittfogel (1957). In the New World, Palerm

(1974) and others have begun to investigate this problem in central Mexico. Similar work in Guatemala has not yet begun.

The preponderant thrust in theory concerning the transformation of peasant ecotypes has been within "development" theory. Brookfield (1975) has traced the rise of this body of theory. The segment of this theory that most concerns this study is the notion of "dual economies." Starting about 1910, Boeke (1953) wrote on the obvious cleavage in Indonesian society. He saw two economies (export and subsistence) and two societies (European-oriented and Asiatic). This was simply a recapitulation of the notion of fundamental differences between East and West (or poor and rich) implicit in European social thought since Malthus, if not earlier. This approach does not address the question of how the West was responsible for creating much of this "dualism."

As has been suggested in this study, Geertz's (1963) revision of Boeke's dual economy thesis is helpful in viewing Guatemala. The idea of "dual nations" neatly fits the Guatemalan situation. Beyond this, Geertz's model is only partially applicable. Other theories must be consulted. As stressed in places in this work, Wallerstein's (1974) study of the sixteenth century and the making of the "modern world system" offers an excellent global perspective on the transformation of peasant ecotypes. In addition, Wallerstein's approach is fundamentally geographical. It is a signal contribution to the study of "core/periphery" relationships. Unfortunately, his theory does not address the question of how precursor "core/periphery" relationships were formed and articulated. The nature of pre-sixteenth century peasant ecotypes in Guatemala remains little understood.

Among geographers, Brookfield (1968, 1972, 1973, 1975) has spoken most cogently about the directions the study of peasant ecotypes and their transformations might take. He is concerned with systems as they exist today, but within the light of reconstructions of the past. Without an understanding of the agronomic practices and strategies for agricultural production in the pre-Conquest past, one can hardly hope to understand the nature of the changes that have produced the distinctive (though not unique) systems of intensive irrigated horticulture in highland Guatemala. Brookfield (1973:10) remarked: "We thus became microgeographers in order that we might be better macrogeographers." One might add, echoing Carl O. Sauer, that we should become historical geographers so that we might be better geographers of the present.

New Practical Directions

As suggested in the preface to this study, the "social empirical" perspective offers both a tradition and a basis for future work in the cultural and historical study of traditional agriculture. At the same time, an important body of experimental as well as empirical data is emerging based on the efforts of workers in several disciplines to test various systems of traditional intensive agriculture. Experiments by botantists, agro-ecologists, archaeologists, and geographers in reconstructing ancient systems for a variety of purposes offers a new dimension to "social empirical" research.

The initial work by Puleston (1977b) in Belize and Gómez-Pompa and his associates in Mexico (1978) was related to questions involving the nature of specific Maya subsistence systems, both ancient and contemporary. Work done more recently, especially in coastal Mexico, has been oriented toward demonstrating the efficacy of combining raised-field farming with aquaculture. This has involved considerable improvisation. The presumed techniques of antiquity have been used as a starting point, but rather than recreate pre-Columbian systems for strictly academic purposes, the emphasis has been on developing alternatives to the current land use practices (Gómez-Pompa et al. 1982). In addition, archaeologists and geographers have begun to use experimental simulations of aspects of the construction and production of ancient systems, with the object of better understanding the labor inputs and other variables involved in the building and maintenance of various systems (Denevan et al. 1981; Erickson 1982). Most of these efforts have been undertaken in areas where evidence of prehistoric intensive agriculture has been recently uncovered. Therefore, they serve, to some extent, as demonstrations for the feasibility of rehabilitating ancient systems in situ.

There is less reason, of course, to do experimental work in areas where traditional agro-ecosystems are currently in production. For example, in the case of the tablón systems of highland Guatemala, one might wish to test tablón construction, maintenance and production with only the tools, fertilizers, and crops that were available to pre-Columbian farmers. This would be of interest to archaeologists, historical demographers, and historical geographers. However, this would not tell us much more than is already observable about the contemporary tablón system.

Tablón culture represents a highly evolved form of traditional intensive agriculture. As such, it may offer a model to be emulated and introduced into areas sharing social and environmental similarities, but for many reasons, do not have populations that farm as intensively as their Maya counterparts in Guatemala. It would seem that densely populated sectors of several of Guatemala's Central American neighbor nations, as well as many parts of upland Mexico, could profit by the introduction of locally adapted variants of the tablón system. The same might be said for many parts of highland South America, both in the Andean nations and in the upland shield areas of eastern South America.

Even more expansively, the same could be argued for zones within the tropical and semi-tropical uplands of the Old World. There has not been nearly enough inventory and descriptive work by students of agricultural landforms to make broad generalizations about the locations, extent, and nature of tablón-like raised-bed horticulture in Africa and Asia. However, enough is known to suggest that similar systems do exist (Denevan and Turner 1974). Moreover, there is probably a great potential not only for introducing Maya-type tablón farming methods to many areas in the Old World, but also there are undoubtedly aspects of other raised-bed gardening traditions that would be useful to the Maya hill and valley farmers of Guatemala.

Rather than envisioning the only avenue of change open to traditional farmers as a forced-march, temporally determined trajectory toward agro-modernity, at times under a literal gun or more often under its economic equivalent, we might begin to imagine a lateral alternative. Traditional farmers could profit by exchanging knowledge with their Pan-tropical counterparts on a direct basis. Over the past two decades there has been a growing awareness among specialists investigating traditional modes of resources management that peasant practitioners themselves are important repositories of knowledge of local ecosystems and their management. Yet within this emergent climate of respect--even deference--for folkways and wisdom, the flow of information is mediated and monitored in an hierarchial fashion from the field(s) to the global metropoles and their outposts. Only rarely does this information, once "processed," travel back to the places where it was generated.

Direct peasant-to-peasant interchanges of information and expertise are almost non-existent. This would seem to be the next step in the challenge to not only preserve, but to perfect traditional approaches to resource management. In this light, the tablón farmers of Panajachel and elsewhere in highland Guatemala are master agricultural craftsmen. Their expertise, so magisterially registered within the landscapes they modify, should not only be studied, understood, and protected, but, if possible, expanded....even exported.

NOTES

[1] Wolf divides ecotypes into two categories: paleotechnic and neotechnic. Both categories use human or animal labor; but neotechnic systems increasingly use the energy of combustible fuels and the skills of science. The paleotechnic ecotypes are: (1) long-term fallowing (swidden) systems; (2) sectorial fallowing (barbecho) systems; (3) short-term fallowing systems (Eurasian grain farming); (4) permanent cultivation (hydraulic systems);

(5) permanent cultivation of favored plots (infield-outfield systems). Neotechnic ecotypes are: (1) specialized horticulture; (2) dairy farming; (3) balanced livestock and crop raising ("mixed farming"); and (4) tropical commodity crop production.

[2] By "involution" Geertz meant several things. Involution at the level of the individual terrace or sawah means that "the output can be almost indefinitely increased by more careful, fine-comb cultivation techniques (1963:35)." At the national level, agricultural involution refers to the creation of a "late Gothic" quality in the rural agricultural economy of Indonesia. Beginning in the nineteenth century "the tenure systems grew more intricate; tenancy relationships more complicated; cooperative labor arrangements more complex--all in an effort to provide everyone with some niche, however small, in the overall system (1963:82)."

[3] The development of tablón culture as a sub-ecotype can be examined on five different levels: (1) the individual tablón plot; (2) local, or Panajachel and its immediate environs; (3) regional, or the greater Atitlán basin area; (4) national, or Guatemala; and (5) global. At each level there is a specific interplay of ecotypes. At the individual tablón level, specialized horticulture has been superimposed on a hydraulic base. At the local level, six ecotypes are fused into one complex though specialized horticulture. New World hydraulic cultivation and tropical commodity crop (coffee) production are the predominant ecotypes. The ordination is borrowed from Adams' (1975:213) levels of integration in his "model of the evolution of power."

[4] The legacy of this arrangement has roots in the pre-Conquest era. At varying times central Mexican states or empires exercised hegemony over parts of Guatemala (Carmack 1970; Chapman 1957). For example, the cacao coast of Soconusco had been integrated into the Aztec Empire by the 1470s. A distinction should be made between imperial systems and "world systems" as Wallerstein has recently done (1974). The Spanish conquest in the sixteenth century marked the present integration of Inner and Outer Guatemala into not only the Spanish imperial system, but, more accurately, into the "Modern World System" which Spain was instrumental in bringing about.

[5] Parsimony appears to be the guiding principle in establishing distance between individual plants. In this regard, what we might term the "parsimony principle" of tablón intensification is the spatial correlative of labor intensification. It can be directly compared to similar properties in the sawah system as described by Geertz (1963:35). In both systems (sawah and tablón), additional labor and refinement of techniques can be absorbed almost

indefinitely. With tablones, the rearrangement and tighter "packing" of individual plants can be carried out almost indefinitely. The most "space-economic" or spatially efficient plants of their respective classes (whether it be pulses, tubers, squashes, or coles) are the plants grown with the greatest frequency. It is not surprising that Panajachel's specialities are the alliaceous crops: onions and garlic. Of all possible vegetables, they are probably the most space-economic. Moreover, it follows that sweet potatoes, arracacha, peanuts, and other potentially solar obstructive plants are grown on the sides of tablones, and not on top.

One might hypothesize that as the Atitlán area tablón farmers intensity the system (under whatever pressures) they will move progressively toward an onion/garlic orientation. They will exclude those cultigens that are the most "space-dyseconomic."

113

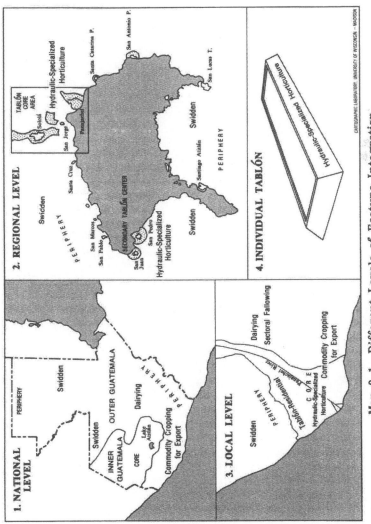

Map 9.1. Different Levels of Ecotype Integration.

Glossary

Azadón: Large broad hoe used for constructing tablones and working milpas.

Brujería: Witchery applied to agricultural situations as well as more general purposes. Carries the connotation of malevolent magic.

Cacaotales: Pre-Conquest cacao plantations, often irrigated.

Canícula (Literally, Dog Star) or Veranillo de San Juan: Short, relatively dry period of varying duration and irregular occurrence, sometime during July/August, between the two periods of maximum rainfall (June and September).

Carga: (Literally) a load or burden. A measure of maize production; a general measure of agricultural production for certain crops: cabbage; avocados, onions: or other produce transported by red.

Chinampas: Extremely productive raised gardens generally fertilized by matter taken from the associated irrigation ditches. Chinampas often combine aquaculture with horticulture in a single complex. They are found primarily in the Valley of Mexico and date to at least A.D. 1000, and probably earlier.

Compadre: Co-father. The relationship of biological father and godfather in the compadrazgo system.

Costumbre: Raison d'etre; the traditional way of doing things; a specific rite or ritual performed by a zahorín or shaman.

Cuerda: A land measurement; usually 32 varas (of 33 inches each) square, which is 7,744 square feet, or between one fifth and one sixth of an acre (43,560 square feet).

Encomendero: Proprietor of an encomienda.

Encomienda: A kind of trust, whereby the crown granted to a conquistador the right to the tributes of a native community, in
exchange for benefits, such as the support of a priest, in-
doctrination, and so on.

Ganado: Livestock.

Huerta: Garden plot.

Ilamo: Almus acuminata, alder tree useful for fuel and lumber.

Juece de Milpa: Spanish official charged with supervising Indian
agriculture.

Ladino: A non-Indian. A class of persons speaking Spanish, who
dress in European-style clothing and, in general, evidence
the Spanish cultural tradition more than the indigenous
one(s).

Lata: Square or rectangular metal container used for transport
and measurement.

Mandamientos: A system of enforced labor whereby the rural em-
ployers were given quotas of laborers.

Milpas: (Literally) a corn field. In Mesoamerica the term is used
for (1) the field in which corn is grown in traditional fashion; (2) occasionally refers to corn itself; and (3) as a syn-
onym for swidden farming.

Monte: The territory outside of a town. Rural area, whether
wooded or not. Also, weeds and vegetative rubbish growing
on or around tablones; green manure applied to tablones.

Mozo: Common laborer, field hand.

Palomas: Pigeons.

Palomitas de Monte: Doves.

Patria Chica: Native village, birth place.

Pensiones: Tourist hotels.

Produit Moteur: (Literally) "Driving Commodity." A product or commodity that was searched for by sixteenth century Europeans which would enable them to gain economic vantage vis à vis their rivals. Gold and silver were the two most sought after commodities, but many lesser products were sought after also.

Pueblo: Town. The smallest category--followed by villa and ciudad--of community that is the seat of government of a municipio. In Panajachel, applied to the whole delta portion as opposed to the monte districts.

Quetzál (Q): Guatemala's national unit of currency pegged to the value of, and equal to, the U.S. dollar.

Reconquista: Campaign (ca. 1100-1492) to drive the Moors from Spain.

Red: A mesh bag made of maguey fiber used to store and transport produce.

Repartimiento: A system of Indian draft labor on a rotating, quota basis. Also forced sales or purchases imposed on Indians or any of the other poorer classes.

Sawah: A flooded paddy field used for grown rice and other foods in Indonesia.

Semillero: Small plots, either at one end of a tablón or housed in a raised rock bed, for starting vegetable seeds.

Swidden: General term for shifting agriculture; "slash and burn" farming; milpa.

Tablón: A raised garden bed; irrigated and often constructed in conjunction with terraces.

Tarakua: Aboriginal hoe found in South Central Mexico (Michoacan and Guerrero).

Tarea: Stint, a unit of work. Usually the amount of work that can be done in one day.

Tierra caliente: (Literally) "hot country." Region below about 1,000 m elevation. Mean annual temperature above about 22 degrees centigrade.

<u>Tierra fría</u>: (Literally) "cold country." Region above about 2,000 m elevation. Mean annual temperature below about 16 degrees centigrade.

<u>Tierra templada</u>: (Literally) "temperate country." Region between about 1,000 and 2,000 m elevation.

<u>Vara</u>: (Literally) "staff." A linear measurement, 32 or 33 inches, as specified.

<u>Zahorín</u>: A diviner. A person versed in acts of witchery and their countermeasures.

Bibliography

Acosta Saignes, M.
   1945   "Les Pochteca." Acta Antropológica 1(1).

Adams, Richard N.
   1957   Cultural Surveys of Panama, Nicaragua, Guatemala, El
           Salvador, Honduras. Washington: Pan American Sanitary
           Bureau, Scientific Publications 33.

   1965   Migraciones Internas en Guatemala: Expansión Agraria
           de los Indígenas Kekchíes hacia El Petén. Guatemala:
           Seminário de Integración Social Guatemalteca, Estudios
           Centroamericanos 1.

   1967a   "Nationalization." In Social Anthropology, Manning Nash
           (ed.). In Handbook of Middle American Indians
           (6:469-489), Robert Wauchope (ed.). Austin: University
           of Texas Press.

   1967b   The Second Sowing: Power and Secondary Development
           in Latin America. Austin: University of Texas Press.

   1970   Crucifixion by Power: Essays on Guatemalan National
           Social Structure, 1944-1966. Austin: University of
           Texas Press.

   1975   Energy and Structure: A Theory of Social Power.
           Austin: University of Texas Press.

Altee, Charles B., Jr.
   1968   Vegetable Production in Guatemala. Guatemala:
           U.S.A.I.D.

120

Alvarado, Pedro de
1924 An Account of the Conquest of Guatemala, in 1524.
      Sedley J. Mackie (ed.), New York: The Cortes Society.

Armillas, Pedro
1971 "Gardens on Swamps." Science 174:653-661.

Arriola, Jorge L.
1973 El Libro de las Geonimias de Guatemala. Guatemala:
      Seminario de Integración Social Guatemalteca.

Bancroft, Hubert H.
1976 The Native Races of the Pacific States. New York: D.
      Appleton & Co.

Benedict, Francis G. and M. Steggerda
1937 "The Food of the Present Day Maya Indians of Yuca-
      tán." Contributions to American Anthropology and His-
      tory 3:155-188. Washington, D.C.: Carnegie Institution
      of Washington, Publication 456.

Bennett, Charles R.
1962 "The Bayano Cuna Indians, Panama: An Ecological
      Study of Livelihood and Diet." Annals of the Association
      of American Geographers 54:32-50.

1964 "Stingless Bee Keeping in Western Mexico." The Geo-
      graphical Review 54:85-92.

Bergman, Roland W.
1969 Shifting Cultivation in the High Rainforest: The
      Chirripo Indians, Costa Rica. M.S. Thesis, Department
      of Geography, University of Wisconsin, Madison.

Bergmann, John F.
1969 "The Distribution of Cacao Cultivation in Pre-Columbian
      America." Annals of the Association of American Geo-
      graphers 59:85-96.

Betancor, A. P. and Pedro de Arboleda
1964a "Descripción de San Bartolomé del Partido de Atitlán"
      (1585). Anales de Geografía e Historía de Guatemala
      38:262-276.

1964b "Relación de Santiago Atitlán (1585)." Anales de Geo-
      grafía e Historía de Guatemala 37:87106.

Blom, Franz
1946   "Apuntes sobre los ingenieros Mayas." Irrigación en Mexico 27:5-16.

Boeke, J. H.
1953   Economics and Economic Policy of Dual Societies. Haarlem: H. D. Tjeenk Willink.

Boserup, Esther
1965   The Conditions of Agricultural Growth. Chicago: Aldine.

Bowen, Margarita
1981   Empiricism and Geographic Thought from Francis Bacon to Alexander von Humboldt. Cambridge: University of Cambridge Press.

Brand, Donald D.
1951   Quiroga: A Mexican Municipio. Washington, D.C.: Smithsonian Institution, Institute of Social Anthropology, Publication 11.

Brigham, William T.
1887   Guatemala: The Land of the Quetzal. Gainesville: University of Florida Press, reprinted 1965.

Brinton, Daniel D.
1885   The Annals of the Cakchiquels. Philadelphia.

Bronson, Bennet
1966   "Roots and the Subsistence of the Ancient Maya." Southwestern Journal of Anthropology 22:251-279.

Brookfield, Harold C.
1968   "New Directions in the Study of Agricultural Systems." In Evolution and Environment (pp. 413-439), E. T. Drake (ed.). New Haven: Yale University Press.

1972   "Intensification and Disintensification in Pacific Agriculture: A Theoretical Approach." Pacific Viewpoint 13:30-48.

1973a   "Introduction: Explaining or Understanding?" In The Pacific in Transition (pp. 3-23), H. C. Brookfield (ed.). New York: St. Martin's Press.

1973b   "Full Circle in Chimbu: A Study of Trends and Cycles." In The Pacific in Transition (pp. 127-160). H. C. Brookfield (ed.). New York: St. Martin's Press.

1975   <u>Interdependent Development</u>. Pittsburgh: University of Pittsburgh Press.

Brush, Stephen B.
1976   "Man's Use of an Andean Ecosystem." <u>Human Ecology</u> 4:147-166.

Bunzel, Ruth
1953   <u>Chichicastenango: A Guatemalan Village</u>. Locust Valley: J. J. Augustin. American Ethnological Society Publications 22.

Butzer, Karl
1976   <u>Early Hydraulic Civilization in Egypt</u>. Chicago: University of Chicago Press.

Cancian, Frank
1967   "Political and Religious Organizations." In <u>Handbook of Middle American Indians</u> (6:283-298), R. Wauchope (ed.). Austin: University of Texas Press.

Cardoso, C. F. S.
1975   "Colonial Modes of Production." <u>Critique of Anthropology</u> 4:1-36.

Carmack, Robert M.
1968   <u>Toltec Influence on the Post-classic Culture History of Highland Guatemala</u>. New Orleans: Middle America Research Institute, Publication 26.

1973   <u>Quichean Civilization: The Ethnohistoric, Ethnographic, and Archeological Sources</u>. Berkeley: University of California Press.

_____ and D. T. Wallace (eds.)
1977   <u>Archaeology and Ethnohistory of the Central Quiché</u>. Albany: The Institute for Mesoamerican Studies.

Carter, George F.
1971   "Pre-Columbian Chickens in America." In <u>Man Across the Sea</u> (pp. 178-218), Carol L. Riley et al. (eds.). Austin: University of Texas Press.

Carter, William E.
1969   <u>New Lands and Old Traditions: Kekchi Cultivators in the Guatemalan Lowlands</u>. Gainesville: University of Florida Press.

Chapman, Anne
1957　"Port of Trade Enclaves in Aztec and Maya Civilization." In Trade and Markets in Early Empires (pp. 114-153), K. Polanyi et al., (eds.). Glencoe: Free Press.

Clarke, William C.
1971　Place and People: An Ecology of a New Guinean Community. Berkeley: University of California Press.

1977　"The Structure of Permanence: The Relevance of Self-substance Communities for World Ecosystem Management." In Subsistence and Survival (pp. 363-384), T. P. Bayliss-Smith (ed.). London: Academic Press.

Coe, Michael D.
1962　"The Chinampas of Mexico." Scientific American 221:90-98.

Conklin, Harold C.
1963　The Study of Shifting Cultivation. Washington, D.C.: Pan American Union, Studies and Monographs 11.

Cortes y Larraz, Pedro
1958　"Descripción Geográfico-Moral de la Diócesis de Guatemala." Biblioteca "Goathemala." Guatemala: Sociedad de Geografiá e Historía de Guatemala, Vol. 22.

Cowgill. Ursula M.
1971　"Some Comments on Manihot Subsistence and the Ancient Maya." Southwestern Journal of Anthropology 27:51-63.

Curry-Lindahl, Kai
1972　Let Them Live. New York: Morrow.

Dahlberg, K. A.
1979　Beyond the Green Revolution. New York: Plenum Press.

Dahlin, Bruce H.
1979　"Cropping Cash in the Protoclassic: A Cultural Impact Statement." In Maya Archaeology and Ethnohistory (pp. 21-37), Norman Hammond and Gordon R. Willey (eds.). Austin: University of Texas Press.

Denevan, William M.
1966   "A Cultural-Ecological View of the Former Aboriginal Settlement in the Amazon Basin." Professional Geographer 18:346-351.

1976a   "The Aboriginal Population of Amazonia." In The Native Population of the Americas in 1492 (pp. 205-233), W. M. Denevan (ed.). Madison: University of Wisconsin Press.

1976b   The Native Population of the Americas in 1492. Madison: University of Wisconsin Press.

1980a   "Latin America." In World Systems of Traditional Resource Management (pp. 217-244), Gary A. Klee (ed.). New York: John Wiley & Sons.

1980b   "Recent Research on Traditional Food Production in Latin America." In Geographic Research on Latin America: Benchmark 1980 (pp. 176-187), T. L. Martinson and G. S. Elbow (eds.). Muncie: Conference of Latin Americanist Geographers.

1980c   "Tipología de Configuraciones Agrícolas Prehispánicas." América Indígena 40:619-652.

1982   "Hydraulic Agriculture in the American Tropics: Forms, Measures, and Recent Research." In Maya Subsistence (pp. 181-203), Kent V. Flannery (ed.). New York: Academic Press.

_____, Kent Mathewson and Richard G. Whitten
1981   "Mounding, Mucking and Mangling: Recent Research on the Raised Fields in the Guayas Basin, Ecuador." Paper read at the Conference on Prehistoric Intensive Agriculture in the Tropics, Australian National University, Canberra.

_____ and B. L. Turner, II
1974   "Forms, Functions and Associations of Raised Fields in the Old World Tropics." Journal of Tropical Geography 39:24-33.

Donkin, Robin A.
1970   "Pre-Columbian Field Implements and Their Distribution in the Highlands of Middle and South America." Anthropos 65:505-529.

1979 Agricultural Terracing in the New World. Tucson: Viking Fund Publications in Anthropology 56.

Downing, Theodore E. and M. Gibson (eds.)
1975 Irrigation's Impact on Society. Tucson: University of Arizona Press.

Dunbar, Gary S.
1974 "Geographical Personality." In Man and Cultural Heritage (pp. 25-33). H. J. Walker and W. G. Haag (eds.). Baton Rouge: Louisiana State University Press.

Erickson, Clark L.
1982 Experiments in Raised Field Agriculture, Huatta, Peru 1981-82. Unpublished manuscript.

Foster, George M.
1960 Culture and Conquest. New York: Wenner-Gren Foundation.

Fuentes y Cuzman, F. A. de
1932- "Recordación Florida, Guatemala, Discurso historical y
1933 demonstración natural, material, militar y política del Reyno de Guatemala." Biblioteca "Goathemala." Guatemala: Sociedad de Geografía e Historía, Vols. 6-8.

Furst, Peter R.
1974 "Morning Glory and Mother Goddess at Tepantitla, Teotihuacán: Iconography and Analogy in Pre-Columbian Art." In Mesoamerican Archaeology: New Approaches (pp. 187-215), Norman Hammond (ed.). London: Duckworth.

Gade, Daniel
1967 "The Guinea Pig in Andean Folk Culture." Geographical Review 57:213-224.

Gann, Thomas W. F. and J. Eric S. Thompson
1931 The History of the Maya from the Earliest Day to the Present. New York: Charles Scribner's Sons.

Garcia Pelaez, F. de F.
1968 "Memorias para la Historía del Antiguo Reino de Guatemala." Biblioteca "Goathemala." Guatemala: Sociedad de Geografía e Historía, Vol. 29.

Garcia Granados, Rafael
1937 Filias y Fobias. Mexico.

Geertz, Clifford
1963   Agricultural Involution: The Process of Ecological Change in Indonesia. Berkeley: University of California Press.

Gibson, Charles
1952   Tlaxcala in the Sixteenth Century. New Haven: Yale University Press.

Gliessman, S. , R. Garcia and M. Amador
1978   Módulo de Producción Diversificada: Un Agroecosistema de Producción Sostenida para el Trópico Húmedo de México. Cárdenas, Tabasco: Colegio Superior de Agricultura Tropical.

Gómez-Pompa, Arturo
1978   "An Old Answer to the Future." Mazingira (Oxford) 5.

_____ and Raúl Venegas
1976   "La Chinampa Tropical." INIREB Informa (Xalapa) 5.

_____ , H. L. Morales, E. Jímenez and J. Jímenez Avilla
1982   "Experiences in Traditional Hydraulic Agriculture." In Maya Subsistence (pp. 327-342), Kent V. Flannery (ed.). New York: Academic Press.

Guzman, Louis E.
1961   "Las terrazas de los antiguos mayas montaneses Chiapas, Mexico." Revista Interamericana de Ciencias Sociales 1(3):398-406.

Halle, Louis J., Jr.
1936   Transcaribbean: A Travel Book of Guatemala, El Salvador, British Honduras. New York.

Harris, Marvin
1968   The Rise of Anthropological Theory. New York: Thomas Crowell.

1974   Cows, Pigs, Wars and Witches. New York: Vintage.

Hill, A. David
1964   The Changing Landscape of a Mexican Municipio-Villa las Rosas, Chiapas. Chicago: Department of Geography, University of Chicago, Research Paper 91.

Hinshaw, Robert
  1968   "Panajachel." In Los Pueblos del Lago de Atitlán. Gua-
         temala: Seminario de Integración Social Guatemalteca,
         Publication 23:69-92.

  1975   Panajachel: A Guatemalan Town in Thirty-year Perspec-
         tive. Pittsburgh: University of Pittsburgh Press.

Humboldt, Alexander von
  1966   Ensayo Político Sobre el Reino de la Nueva España.
         Mexico: Editorial Porrua.

Huxley. Aldous
  1934   Beyond the Mexique Bay. New York: Harper Brothers.

Jonas, Susanne
  1974   "Guatemala: Land of Eternal Struggle." In Latin Ameri-
         ca: The Struggle with Dependency and Beyond (pp.
         89-219), R. Chilcote and J. Edelstein (eds.). New
         York: John Wiley & Sons.

Jones, Chester L.
  1940   Guatemala: Past and Present. Minneapolis: University of
         Minnesota Press.

Juarros, Domingo
  1823   A Statistical and Commercial History of the Kingdom of
         Guatemala in Spanish America. London.

Kirchhoff, Paul
  1952   "Meso-America." In Heritage of Conquest (pp. 17-30),
         Sol Tax (ed.). New York: MacMillan.

Kropotkin, Peter
  1902   Mutual Aid: A Factor in Evolution. London: Heinemann.

Las Casas, Bartolome de
  1967   Apologética Historía Sumaria. Mexico

Lothrop, Samuel K.
  1928   "Santiago Atitlán, Guatemala." Indian Notes 5:370-395.

  1929   "Canoes of Lake Atitlán." Indian Notes 6:216-222.

  1933   Atitlán: An Archeological Study of Ancient Remains on
         the borders of Lake Atitlán. Washington, D.C.:
         Carnegie Institution of Washington, Publication 444.

128

Lovell, George W.
  1981 "The Historical Demography of the Cuchumatan High-
          lands of Guatemala, 1500-1821." In _Studies in Spanish_
          American Population History (pp. 196-203), David J.
          Robinson (ed.). Boulder: Westview Press.

  1983 "To Submit and to Serve: Forced Native Labour in the
          Cuchumatán Highlands of Guatemala; 1525-1821." _Journal_
          of Historical Geography 9(2):127-144.

Lundell, Cyrus L.
  1938 "The Agriculture of the Maya." _Southwest Review_
          19:65-77.

Lutz, Christopher
  1976 _Santiago de Guatemala, 1541-1773: The Socio-Demogra-_
          phic History of a Spanish American Colonial City. Ph.D.
          dissertation, Department of History, University of
          Wisconsin, Madison.

Mackie, Sedley J. (ed.)
  1924 _An Account of the Conquest of Guatemala in 1524 by_
          Pedro Alvarado. New York: The Cortes Society.

MacLeod. Murdo J.
  1973 _Spanish Central America: A Socio-Economic History_
          _1520-1720._ Berkeley: University of California Press.

MacNutt, Francis A. (ed.)
  1908 _Letters of Cortés._ Vol. I. New York: C. P. Putnam's
          Sons.

Marcus, Joyce
  1982 "The Plant World of the Sixteenth- and Seven-
          teenth-Century Lowland Maya." In _Maya Subsistence_
          (pp. 239-273), Kent V. Flannery (ed.). New York: Ac-
          ademic Press.

Martinez, Maximino
  1928 _Plantas Utiles de la Republica Mexicana._ Mexico.

Matheny, Ray
  1976 "Maya Lowland Hydraulic Systems." _Science_ 193:639-646.

          and D. Gurr
  1979 "Ancient Hydraulic Techniques in the Chiapas High-
          lands." _American Scientist_ 67:441-449.

Mathewson, Kent
   1976   Specialized Horticulture in the Guatemalan Highlands:
          The Tablón System of Panajachel. M.A. thesis,
          Department of Geography, University of Wisconsin,
          Madison.

   1977   "Maya Urban Genesis Reconsidered: Trade and Intensive
          Agriculture as Primary Factors." Journal of Historical
          Geography 3:203-215.

   1978   "Tablón Culture: The Question of Origins." Paper pre-
          sented at the 68th Annual Meeting of the Association of
          American Geographers, New Orleans.

   1979   "Agricultural Intensity Measurement: The Case of Crop
          Terracing in Highland Guatemala." Paper presented at
          the 69th Annual Meeting of the Association of American
          Geographers, Philadelphia.

   1983   "Taxonomy of Raised and Drained Fields: A Morpho-
          genetic Approach." In Prehistoric Intensive Agriculture
          in the Tropics, Ian S. Farrington (ed.). Oxford:
          British Archaeological Reports, in press.

Maudslay. A. C. and A. P. Maudslay
   1899   A Glimpse at Guatemala and Some Notes on the Ancient
          Monuments of Central America. London.

McBryde, Felix W.
   1933   Sololá: A Guatemalan Town and Cakchiquel Market Cen-
          ter. New Orleans: Middle American Research Institute,
          Publication 5:45-152.

   1940   "Influenza in America During the Sixteenth Century
          (Guatemala, 1523, 1559-62, 1576)." Bulletin of the His-
          tory of Medicine 8:296-302.

   1947   Cultural and Historical Geography of Southwest Guate-
          mala. Washington, D.C.: Smithsonian Institution, Insti-
          tute of Social Anthropology, Publication 4.

McQuown, Norman A.
   1956   "The Classification of the Mayan Languages." Interna-
          tional Journal of American Linguistics 22:191-195.

130

Millon, Rene F.
1955   "Trade, Tree Cultivation and the Development of Private Property in Land." American Anthropology 52:698-712.

Murra, John V.
1972   "El 'control vertical' de un maximo de pisos ecológicos en la economía de las sociedades Andinas." In Visita de la Provincia de León de Huánuco (1562), Iñigo Ortiz de Zuñiga, Visitador (II:429-476). Huánuco, Peru: Universidad Hermillo Valdizan.

Netting, Robert McC.
1977   "Maya Subsistence: Mythologies, Analogies, Possibilities." In The Origins of Maya Civilization (pp. 299-333), R. E. W. Adams (ed.). Albuquerque: University of New Mexico Press.

Nordenskiöld, Erland
1929   "L'Apiculture Indienne." Journal de la Société des Américanistes de Paris, N.S. 21:17.

Nuttal, Zelia
1920   "Los Jardines del Antiguo Mexico." Société Scientifique "Antonio Alzate"-Mémoires 37:193-213.

Ocana, Fr. Diego de
1933   "Descripción de la Laguna de Atitlán." Anales de la Sociedad de Geografía e Historia 9:297-302.

Paez Betancor, A. and P de. Arboleda
1965   "Descripción de San Bartolomé del Partido de Atitlán año 1585." Anales de la Sociedad de Geografía e Historia 38:262-276.

Palerm, Angel
1954   "La distribución del regadío en el área central de Meso-America." Ciencias Sociales. Washington, D.C.: Pan American Union 5:25-26; 2(15):64-74.

1955   "The Agricultural Base of Urban Civilization in Mesoamerica." Social Science Monographs. Washington, D.C.: Pan American Union 1:28-42.

1967   "Agricultural Systems and Food Patterns." In Social Anthropology, Manning Nash (ed.). In Handbook of Middle American Indians (6:26-52), Robert Wauchope (ed.). Austin: University of Texas Press.

1973   Obras Hidraúlicas Prehispánicas. Mexico: Instituto Nacional de Anthropología e Historía.

Paul, Benjamin D.
1968   "San Pedro La Laguna." In Los Pueblos del Lago de Atitlán. Guatemala: Seminario de Integración Social Guatemalteca, Publicacion 23:93-158.

Polanyi, Karl
1957   Trade and Market in the Early Empires. Glencoe: Free Press.

Puleston, Dennis E.
1968   Brosimum alicastrum as A Subsistence Alternative for the Classic Maya of the Central Southern Lowlands. M.A. Thesis, Department of Anthropology, University of Pennsylvania, Philadelphia.

1971   "An Experimental Approach to the Function of Maya Chultuns." American Antiquities 36:322- 335.

1976   Personal Communication

1977a   "The Art and Archaeology of Hydraulic Agriculture in the Maya Lowlands." In Social Process in Maya Prehistory (pp. 449-467), Norman Hammond (ed.). New York: Academic Press.

1977b   "Experiments in Prehistoric Raised Field Agriculture: Learning from the Past." Journal of Belizean Affairs 5:36-43.

1978   "Terracing, Raised Fields, and Tree Cropping in the Maya Lowlands: A New Perspective on the Geography of Power." In Pre-Hispanic Maya Agriculture (pp. 225-245), P. D. Harrison and B. L. Turner, II (eds.). Albuquerque: University of New Mexico Press.

_____ and John P. Bradbury
1974   "Experimental Agriculture as a Method for Testing Models of Ancient Subsistence Systems." Paper presented at 73rd Annual Meeting of the American Anthropological Association, Mexico City, Mexico.

_____ and Olga S. Puleston
1971   "An Ecological Approach to the Origins of Maya Civilization." Archaeology 24:330-337.

Recinos, Adrian
1954   Monografía de Huehuetenango. Guatemala: Ministerio de Educación Publica.

Remesal, A. de
1932   "Historía General de las Indias Occidentales, y Particular de la Gobernación de Chiapa y Guatemala." Biblioteca "Goathemala." Guatemala: Sociedad de Geografía e Historía de Guatemala, Vol. 4-5.

Robicsek, Frances
1978   The Smoking Gods: Tobacco in Maya Art, History and Religion. Norman: University of Oklahoma Press.

Rojas-Lima, Flavio
1968   "Otros pueblos del lago." In Los Pueblos del Lago de Atitlán. Guatemala: Seminario de Integración Social Guatemalteca, Publicación 23:277-340.

Roys, Ralph L.
1931   "The Ethnobotany of the Maya." Middle American Research Series 2. New Orleans: Tulane University.

Sanders, William T. and Carson Murdy
1982   "Population and Agricultural Adaptation in the Humid Highlands of Guatemala." In The Historical Demography of Highland Guatemala (pp. 23-34), R. M. Carmark, J. Early, and C. Lutz (eds.). Albany: The Institute for Mesoamerican Studies.

Sapper, Karl
1897   Sobre la Geografía Física, la Populación y la Producción de la Republica de Guatemala. Guatemala.

1935   "Bienenhaltung und Bienenzucht in Mittelamerika und Mexico." Ibero-Amerikanisches Archiv. 9:183-198.

Sauer, Carl O.
1936   "American Agricultural Origins: A Consideration of Nature and Culture." In Essays in Anthropology Presented to A.L. Kroeber in Celebration of His Sixtieth Birthday, June 11, 1936 (pp. 279-297). Berkeley: University of California Press.

1941   "The Personality of Mexico." Geographical Review 31: 353-364.

1952   Agricultural Origins and Dispersals. New York: American Geographical Society.

1958   "Man in the Ecology of Tropical America." Proceedings of the Ninth Pacific Science Congress, Bangkok, 1957. 20:104-110.

1959   "Age and Area of American Cultivated Plants." Actas 1:215-229. 33rd. Congreso Internacional de Americanistas, San Jose, Costa Rica.

1981   "Indian Food Production in the Caribbean." The Geographical Review 71:272-280.

Serpenti, L. M.
1965   Cultivators of the Swamps - Social Structure and Horticulture in a New Guinea Society. Assen: Royal Van Gorlum.

Schmid, Lester
1967   Migrant Labor on the Pacific Coast of Guatemala. Ph.D. dissertation, University of Wisconsin, Madison.

1969   The Development of Family Farms in Guatemala. Unpublished manuscript. Land Tenure Center, University of Wisconsin, Madison.

Schultze-Jena, L.
1945   "La vida y las creencias de los indios quiches de Guatemala." Anales de la Sociedad de Geografía e Historia de Guatemala 20:145-160.

Sherman, William L.
1979   Forced Native Labor in Sixteenth Century Central America. Lincoln: University of Nebraska Press.

Siemens, Alfred H. and Dennis E. Puleston
1972   "Ridged Fields and Associated Features in Southern Campeche: New Perspectives on the Lowland Maya." American Antiquity 37:228-239.

Smith, A. Leyard
1955   Archaeological Reconnaissance in Central Guatemala. Washington, D.C.: Carnegie Institute of Washington, Publication 608.

Smith, Carol A.
　1972　"Market Articulation and Economic Stratification in Western Guatemala." Food Research Institute Studies 11:203-233.

　1976　"Causes and Consequences of Central-Place Types in Western Guatemala." In Regional Analysis, Vol. I, Economic Systems (pp. 255-300), C. A. Smith (ed.). New York: Academic Press.

　1977　"How Marketing Systems Affect Economic Opportunity in Agrarian Societies." In Peasant Livelihood (pp. 117-146), R. Halperin and J. Dow (eds.). New York: St. Martin's Press.

Soustelle, Jacques
　1961　Daily Life of the Aztecs. Stanford: Stanford University Press

Spencer, Joseph E. and G. A. Hale
　1961　"The Origin, Nature, and Distribution of Agricultural Terracing." Pacific Viewpoint 2:1-40.

Squire, Ephriam George
　1858　The States of Central America. New York: Hayser and Brothers.

Stadelman, Raymond
　1940　"Maize Cultivation in Northwestern Guatemala." Contributions to American Anthropology and History. Washington, D.C.: Carnegie Institute Publication 523, pp. 83-263.

Standley, Paul C.
　1930　Flora of Yucatán. Chicago: Field Museum of Natural History, Botanical Series 32(3), Publication 279.

Stanislawski, Dan
　1947　"Early Spanish Town Planning in the New World." Geographical Review 37:94-105.

Stanley, Sam
　1975　"The Panajachel Symposium." Current Anthropology 16:518-523.

Steward, Julian H.
　1955　The Theory of Cultural Change. Urbana: University of Illinois Press.

Steward, Julian H.
1955   Irrigation Civilizations: A Comparative Study. Social Science Monographs I. Washington, D.C.: Pan American Union.

Tax, Sol
1937   "The Municipios of the Midwestern Highlands of Guatemala." American Anthropologist 39:

1946   "The Towns of Lake Atitlán." University of Chicago, Microfilm Collection. MMS of Middle American Cultural Anthropology 13.

1950   "Panajachel: Field notes." University of Chicago, Microfilm Collection. MMS of Middle American Cultural Anthropology 29.

1953   Penny Capitalism: A Guatemalan Indian Economy. Washington, D.C.: Smithsonian Institution, Institute of Social Anthropology, Publication 16.

1975   "The Bow and the Hoe." Current Anthropology 16:507-513.

_____ and Robert Hinshaw
1969   "The Maya of the Midwestern Highlands." In Ethnology, Part I, E. Z. Vogt (ed.). In Handbook of Middle American Indians (7:69-100), Robert Wauchope (ed.). Austin: University of Texas Press.

Termer, Franz
1957   Etnología y Etnografía de Guatemala. Guatemala: Seminario de Integración Social Guatemaltecoa, Publication 5.

Thompson. J. Eric S.
1964   "Trade Relations Between the Maya Highlands and Lowlands." Estudios de Cultura Maya 4:13-49.

1970   Maya History and Religion. Norman: University of Oklahoma Press.

Turner, B. L., II
1974a   Prehistoric Intensive Agriculture in the Mayan Lowlands: New Evidence From the Río Bec Region. Ph.D. dissertation, Department of Geography, University of Wisconsin, Madison.

1974b Prehistoric Intensive Agriculture in the Mayan Lowlands." Science 185:118-124.

1976 "Prehistoric Population Density in the Maya Lowlands: New Evidence for Old Approaches." The Geographical Review 66:73-82.

1978 "The Development and Demise of the Swidden Thesis of Maya Agriculture." In Pre-Hispanic Maya Agriculture (pp. 13-22), Peter D. Harrison and B. L. Turner, II (eds.). Albuquerque: University of New Mexico Press.

1979 "Prehispanic Terracing in the Central Maya Lowlands: Problems of Agricultural Intensification." In Maya Archaeology and Ethnohistory (pp. 103-115), Norman Hammond and Gordon R. Willey (eds.). Austin: University of Texas Press.

1983 Once Beneath the Forest: Prehistoric Terracing in the Río Bec Region of the Maya Lowlands. Boulder: Westview Press, Dellplain Latin American Studies 13.

Vasquez, R. P. Fr. Francisco
1944 "Crónica de la Provincia del Santísimo Nombre de Jesús de Guatemala." Bibliografía "Goathemala." Guatemala: Sociedad de Geografía e Historia de Guatemala, Vol. 17(4).

Veblen, Thomas T.
1974 The Ecological, Cultural and Historical Bases for Forest Preservation in Totonicapán, Guatemala. Ph.D. dissertation, Department of Geography, University of California, Berkeley.

1977 "Native Population Decline in Totonicapán, Guatemala." Annals of the Association of American Geographers 67(3):484-499.

Wallerstein, Immanuel
1974 The Modern World System: Capitalist Agriculture and the Origins of the European World-Economy in the Sixteenth Century. New York: Academic Press.

1975 "Rise and Future Demise of the World Capitalist System." Comparative Studies in Society and History 16:387-415.

West, Robert C.
1948   Cultural geography of the Modern Tarascan Area. Washington, D.C.: Smithsonian Institution, Institute of Social Anthropology, Publication 7.

_____ and Pedro Armillas
1950   "Las Chinampas de Mexico." Cuadernos Americanos 50:165-182.

Wilken, Gene C.
1967   Drained-Field Agriculture in Southwest Tlaxcala, Mexico. Ph.D. dissertation, Department of Geography, University of California, Berkeley.

1969   "Drained-Field Agriculture: An Intensive Farming System in Tlaxcala, Mexico." Geographical Review 59:215-241.

1970   "The Ecology of Gathering in a Mexican Farming Region." Economic Botany 24:286-295.

1971   "Food-Producing Systems Available to the Ancient Maya." American Antiquity 36:438.

1972   "Microclimate Modification by Traditional Farmers." Geographical Review 62:554-560.

1974   Personal Communication.

1977a   "Integrating Forest and Small-Scale Farm Systems in Middle America." Agro-Ecosystems 3:291-302.

1977b   "Manual Irrigation in Middle America." Agricultural Water Management 1:155-165.

1977c   "Minor Agricultural Landforms in Middle America." Paper presented at the Annual Meeting of the Association of American Geographers, Salt Lake City.

1978   "Management of Productive Space in Traditional Farming." Actes du XLIIe Congrès International des Américanistes (Paris) 2:409-419.

1979   "Traditional Slope Management: An Analytical Approach." Hill Lands: Proceedings of an International Symposium (pp. 416-422). Morgantown: West Virginia University Books.

138

Wisdom, Charles
1940   The Chorti Indians of Guatemala. Chicago: University of Chicago Press.

Wittfogel, Karl A.
1957   Oriental Despotism: A Comparative Study of Total Power. New Haven: Yale University Press.

Wolf, Eric R.
1957   "Closed Corporate Peasant Communities in Mesoamerica and Central Java." Southwestern Journal of Anthropology 13:1-18.

1966   Peasants. Englewood Cliffs: Prentice-Hall.

Woods, Clyde R.
1974   Personal Communication.

Ximenez, Francisco
1929-   "Historia de la Provincia de San Vicente de Chiapa y
1931   Guatemala, 1715-1721." Biblioteca "Goathemala." Guatemala: La Sociedad de Geografía e Historia, Vols. 1-3.

Appendix A

Table A.1
Cultivated Tablón Crops

	MAJOR PLANTS
(English Name)	(Scientific; Spanish; and Cakchiquel Names)

1. ONION — _Allium_ spp.; Cebolla; q'os
Varieties: 3
Uses:
  Food: Side dish
  Other: Medicinal, ritual, social, economic
Planting Data:
  Incidence: Widespread*
  Time: Year-round (mostly November and December)
  Location: Side, top of ridges, seed beds & transplants
  Spacing: 1 to 3 inches
  Propagation Method: Seed, transplant, bulb
Probable Origin: Old World

2. GARLIC — _Allium sativum_ L.; Ajo; anʃ
Varieties: 1
Uses:
  Food: Side dish
  Other: Medicinal, ritual, social, economic
Planting Data:
  Incidence: Widespread*
  Time: October through June
  Location: Side, top
  Spacing: 4 inches
  Propagation Method: Peel
Probable Origin: Old World

3. STRAWBERRY — _Fragaria_ spp.; Fresa; fras
Varieties: 1
Uses:
  Food: Snack food
  Other: Economic, ritual, social
Planting Data:
  Incidence: Widespread*
  Time: Year-round
  Location: Top of ridges
  Spacing: 5 inches
  Propagation Method: Seed, transplants
Probable Origin: New World

4. BEANS — _Phaseolus_ spp.; Frijol; k'n q'
Varieties: 3
Uses:
  Food: Main staple, snack food
  Other: Economic, ritual, social
Planting Data:
  Incidence: Widespread*
  Time: Year-round
  Location: Top
  Spacing: 5 inches
  Propagation Method: Seed
Probable Origin: New World

5. MAIZE — _Zea Mays_ L.; Maíz; i' ʃin
Varieties: 3
Uses:
  Food: Main staple, snack food
  Other: Medicinal, ritual, technological, social, economic
Planting Data:
  Incidence: Wide spread*
  Time: May through December
  Location: Top of mound
  Spacing: 35 inches
  Propagation Method: seed
Probable Origin: New World

Table A.1 (cont'd)
Cultivated Tablón Crops

6. CARROT	Daucus curota L.; Zanahoria; zanahor
Varieties:	1
Uses:	
Food:	Side dish
Other:	Economic, ritual, social
Planting Data:	
Incidence:	Occasional**
Time:	Year-round
Location:	Top
Spacing:	2 inches, peripheral (not in tablón proper)
Propagation Method:	Seed
Probable Origin:	Old World

7. CABBAGE	Brassica oleracea L.; Repollo; kulʃ
Varieties:	1
Uses:	
Food: ·	Side dish
Other:	Economic, ritual, social
Planting Data:	
Incidence:	Occasional**
Time:	Year-round (mostly November through May)
Location:	Margins (on top), side
Spacing:	12 inches
Propagation Method:	Seed
Probable Origin:	Old World

8. RADISH	Raphanus satirris L.; Rabano; rabano
Varieties:	1
Uses:	
Food:	Side dish
Other:	Economic, ritual, social
Planting Data:	
Incidence:	Occasional**
Time:	Year-round
Location:	Top
Spacing:	2 inches
Propagation Method:	Seed
Probable Origin:	Old World

9. MELON PEAR	Solanum muricatum Rit.; Pepino; pe'pin
Varieties:	2
Uses:	
Food:	Snack food
Other:	Ritual, economic, social
Planting Data:	
Incidence:	Rare***
Time:	July through May
Location:	Own mounds
Spacing:	47 inches
Propagation Method:	Seed
Probable Origin:	Old World

10. LETTUCE	Lactuca sativa; Lechuga; lechug
Varieties:	3
Uses:	
Food:	Side dish
Other:	Medicinal, ritual, economic, social
Planting Data:	
Incidence:	Occasional**
Time:	Year-round
Location:	Top
Spacing:	Planted in patches; isolated specimens
Propagation Method:	Seed
Probable Origin:	Old World

Table A.1 (cont'd)
Cultivated Tablón Crops

11. PEAS   Pisum sativa L.; Arveja; arveja
Varieties: 1
Uses:
  Food: Side dish
  Other: Economic, ritual, social
Planting Data:
  Incidence: Occasional**
  Time: Year-round
  Location: Top
  Spacing: 4 inches
  Propagation Method: Seed
Probable Origin: Old World

12. BEET   Beta vulgaris L.; Remolacha; remolach
Varieties: 1
Uses:
  Food: Side dish
  Other: Technological, economic, ritual, social
Planting Data:
  Incidence: Occasional**
  Time: Year-round
  Location: Top, side
  Spacing: 5 inches
  Propagation Method: Seed
Probable Origin: Old World

13. TURNIP   Brassica rapa L.; Nabo; nabo
Varieties: 1
Uses:
  Food: Side dish
  Other: Economic, ritual, social
Planting Data:
  Incidence: Occasional**
  Time: Year-round
  Location: Top
  Spacing: 4 inches
  Propagation Method: Seed
Probable Origin: Old World

14. COFFEE   Coffea arabica L.; Café; ka'pe
Varieties: 1
Uses:
  Food: No food value
  Other: Medicinal, ritual, social, economic
Planting Data:
  Incidence: Rare***
  Time: Year-round
  Location: Top
  Spacing: 17 inches
  Propagation Method: Seed
Probable Origin: Old World

15. BRAMBLES   Rubus spp.; Frambuesa; mora
Varieties: 3
Uses:
  Food: Snack food (minor)
  Other: Economic, ritual, social
Planting Data:
  Incidence: Rare***
  Time: Year-round
  Location: Top
  Spacing: 6 inches
  Propagation Method: Seed (transplant)
Probable Origin: Old World

Table A.1 (cont'd)
Cultivated Tablón Crops

16. MANIOC	Manihot esculenta Crantz; Yuca; ts'in
Varieties:	1
Uses:	
Food:	Main food
Other:	Technological, economic, ritual, social
Planting Data:	
Incidence:	Occasional**
Time:	Year-round
Location:	Martins (on top)
Spacing:	18 inches
Propagation Method:	Cutting
Probable Origin:	New World

17. SWEET POTATO	Ipomaea batatas (L.) Lam.; Camote; is
Varieties:	4
Uses:	
Food:	Main food
Other:	Economic, ritual, social
Planting Data:	
Incidence:	Widespread*
Time:	Year-round
Location:	Side
Spacing:	10 inches
Propagation Method:	Slip
Probable Origin:	New World

18. ARRACACHA	Arracacia xanthorrhiza Bauer; Arracacha; arracach
Varieties:	1
Uses:	
Food:	Side dish
Other:	Economic, ritual, social
Planting Data:	
Incidence:	Occasional**
Time:	Year-round
Location:	Side
Spacing:	10 inches
Propagation Method:	Slip
Probable Origin:	New World

19. POTATO	Solanum tuberosum L.; Papa; kaʃ'lan is
Varieties:	2
Uses:	
Food:	Side dish
Other:	Economic, ritual, social
Planting Data:	
Incidence:	Rare***
Time:	Year-round
Location:	Top
Spacing:	6 inches
Propagation Method:	Slip
Probable Origin:	New World

LEGUMES

20. CHIPILIN	Crotalaria longirostrata; Chipilín; mutʃ'
Varieties:	1
Uses:	
Food:	Snack food
Other:	Medicinal; ritual; social; economic
Planting Data:	
Incidence:	Occasional**
Time:	Year-round
Location:	Margins (on top), top
Spacing:	Isolated specimens
Propagation Method:	Seed
Probable Origin:	New World

Table A.1 (cont'd)
Cultivated Tablón Crops

---

21. PEANUT   Arachis hypogaea L.; Maní; mani
Varieties: 1
Uses:
     Food: Snack food
     Other: Economic, ritual, social
Planting Data:
     Incidence: Rare***
     Time: Year-round
     Location: Side
     Spacing: 10 inches
     Propagation Method: Seed
Probable Origin: New World

22. GARBANZO   Cicer arietinum L.; Garbanzo; garbanzo
Varieties: 1
Uses:
     Food: Side dish
     Other: Economic, ritual, social
Planting Data:
     Incidence: Occasional**
     Time: Year-round
     Location: Margins (on top); top
     Spacing: Isolated specimens
     Propagation Method: Seed
Probable Origin: Old World

23. BROAD BEAN   Vicia faba L.; Haba; anſ
Varieties: 1
Uses:
     Food: Side dish
     Other: Economic, ritual, social
Planting Data:
     Incidence: Rare***
     Time: Year-round
     Location: Top
     Spacing: 4 inches
     Propagation Method: Seed
Probable Origin: Old World

24. LENTIL   Lens esculenta Moench; Lenteja; lenteja
Varieties: 1
Uses:
     Food: Side dish
     Other: Economic, ritual, social
Planting Data:
     Incidence: Rare***
     Time: Year-round
     Location: Top
     Spacing: 4 inches
     Propagation Method: Seed
Probable Origin: Old World

GREENS AND POT HERBS

25. AMARANTH   Amaranthus paniculatum L.; Bledo Colorado; tsets
Varieties: 2
Uses:
     Food: Side dish
     Other: Medicinal, economic, ritual
Planting Data:
     Incidence: Widespread*
     Time: Year-round
     Location: Top
     Spacing: 3 inches (planted in patches)
     Propagation Method: Seed
Probable Origin: New World

146

Table A.1 (cont'd)
Cultivated Tablón Crops

26. AMARANTH   Amaranthus caudatus L.; Cola de Zorro
Varieties: 1
Uses:
  Food: Side dish
  Other: Medicinal, economic, ritual
Planting Data:
  Incidence: Occasional**
  Time: Year-round
  Location: Top
  Spacing: 3 inches (planted in patches)
  Propagation Method: Seed
Probable Origin: New World

27. APAZOTE   Chenopodium Nuttallia S.; Apazote; si'k'jex
Varieties: 1
Uses:
  Food: Side dish
  Other: Medicinal, economic, ritual
Planting Data:
  Incidence: Widespread*
  Time: Year-round
  Location: Top
  Spacing: 3 inches (planted in patches)
  Propagation Method: Seed
Probable Origin: New World

28. HIERBA BLANCA   Brassica spp.; Hierba blanca
Varieties: 1
Uses:
  Food: Side dish
  Other: Economic
Planting Data:
  Incidence: Widespread*
  Time: Year-round
  Location: Top
  Spacing: 2 inches (in patches)
  Propagation Method: Seed
Probable Origin: New World

29. MUSTARD   Brassica spp.; Mustáz; ra'q n a'k'al
Varieties: 2
Uses:
  Food: Side dish
  Other: Economic
Planting Data:
  Incidence: Occasional**
  Time: Year-round
  Location: Top
  Spacing: Isolated specimens
  Propagation Method: Seed
Probable Origin: Old World

30. SPINACH   Spinacia oleracea L.; Espinaca
Varieties: 1
Uses:
  Food: Side dish
  Other: Medicinal
Planting Data:
  Incidence: Rare***
  Time: Year-round
  Location: Top
  Spacing: 1.5 inches
  Propagation Method: Seed
Probable Origin: Old World

Table A.1 (cont'd)
Cultivated Tablón Crops

---

31. WATER CRESS — <u>Nasturtium</u> <u>officinale</u> R. Br.; Berro de Fuente
Varieties: — 1
Uses:
  Food: — Snack food
  Other: — Medicinal, economic, social
Planting Data:
  Incidence: — Occasional**
  Time: — Year-round
  Location: — Peripheral (planted in irrigation ditches)
  Spacing: — Planted in patches
  Propagation Method: — Transplants
Probable Origin: — Old World

NIGHTSHADES

32. TOMATO — <u>Lycopersicon</u> <u>esculentum</u> Mill.; Tomate
Varieties: — 2
Uses:
  Food: — Side dish
  Other: — Medicinal; ritual; economic; social
Planting Data:
  Incidence: — Rare***
  Time: — Year-round
  Location: — Top
  Spacing: — Isolated specimens
  Propagation Method: — Seed
Probable Origin: — New World

33. TOMATILLO — <u>Lycopersicon</u> <u>esculentum</u> var.; Tomatillo, tomate de culebra
Varieties: — 1
Uses:
  Food: — Side dish
  Other: — Medicinal, ritual, economic, social
Planting Data:
  Incidence: — Occasional**
  Time: — Year-round
  Location: — Top
  Spacing: — Isolated specimens
  Propagation Method: — Seed
Probable Origin: — New World

34. GROUNDCHERRY — <u>Physalis</u> <u>hypogaea</u>; Miltomate; to'mat
Varieties: — 1
Uses:
  Food: — Side dish
  Other: — Medicinal, ritual, economic, social
Planting Data:
  Incidence: — Occasional**
  Time: — Year-round
  Location: — Top
  Spacing: — Isolated specimens
  Propagation Method: — Seed
Probable Origin: — New World

35. CHILI PEPPER — <u>Capsicum</u> <u>annuum</u> <u>L.</u>; Chile
Varieties: — 10
Uses:
  Food: — Main staple, side dish, condiment
  Other: — Medicinal, ritual, economic, social
Planting Data:
  Incidence: — Occasional**
  Time: — Year-round
  Location: — Margins (on top), top
  Spacing: — 12 inches
  Propagation Method: — Seed
Probable Origin: — New World

148

Table A.1 (cont'd)
Cultivated Tablón Crops

36. HIERBA MORA　　Solarium nigrum; Hierba Mora; max'k'uj
Varieties:　　　　　1
Uses:
　Food:　　　　　Condiment
　Other:　　　　Medicinal, ritual, social
Planting Data:
　Incidence:　　Occasional**
　Time:　　　　Year-round
　Location:　　Top
　Spacing:　　Isolated specimens
　Propagation Method: Seed
Probable Origin:　New World

SQUASHES

37. SQUASH　　　Cucurbita pepo L.; Güicoy
Varieties:　　　　1
Uses:
　Food:　　　　Side dish
　Other:　　　Medicinal, ritual, economic, social, technological
Planting Data:
　Incidence:　Rare***
　Time:　　　Year-round
　Location:　Top, top of mounds (with maize)
　Spacing:　Isolated specimens
　Propagation Method: Seed
Probable Origin:　New World

38. SQUASH　　　Cucurbita moschata Duch.; Ayote; k'un
Varieties:　　　　1
Uses:
　Food:　　　　Side dish
　Other:　　　Medicinal, ritual, economic, social, technological
Planting Data:
　Incidence:　Rare***
　Time:　　　Year-round
　Location:　Top, top of mounds (with maize)
　Spacing:　Isolated specimens
　Propagation Method: Seed
Probable Origin:　New World

39. SQUASH　　　Cucurbita ficifolia Bouche; Chilacayote
Varieties:　　　　1
Uses:
　Food:　　　　Side dish
　Other:　　　Medicinal, ritual, economic, social, technological
Planting Data:
　Incidence:　Rare***
　Time:　　　Year-round
　Location:　Top, top of mound (with maize)
　Spacing:　Isolated specimens
　Propagation Method: Seed
Probable Origin:　New World

40. CUCUMBER　　Cucumis sativus L.; Pepino de ensalada
Varieties:　　　　1
Uses:
　Food:　　　　Side dish
　Other:　　　Medicinal, ritual, economic, social
Planting Data:
　Incidence:　Occasional**
　Time:　　　Year-round
　Location:　Top
　Spacing:　Isolated specimens
　Propagation Method: Seed
Probable Origin:　New World

Table A.1 (cont'd)
Cultivated Tablón Crops

41. CHAYOTE   Sechium edule S.W.; Güisquil; k'i
Varieties: 1
Uses:
  Food: Main staple, side dish
  Other: Medicinal, ritual, economic, social
Planting Data:
  Incidence: Occasional**
  Time: Year-round
  Location: Peripheral (rock arbors, etc.)
  Spacing: Isolated specimens
  Propagation Method: Seed
Probable Origin: Old World

CONDIMENTS AND SPICES

42. CINTULA   Not identified; Cintula; sin'tul
Varieties: 1
Uses:
  Food: None
  Other: Medicinal, ritual, social, economic
Planting Data:
  Incidence: Occasional**
  Time: Year-round
  Location: Top
  Spacing: 1 inch (planted in patches)
  Propagation Method: Transplants
Probable Origin: New World

43. CORIANDER   Coriandrum sativum L.; Culandro
Varieties: 1
Uses:
  Food: Condiment
  Other: Medicinal, ritual, economic
Planting Data:
  Incidence: Occasional**
  Time: Year-round
  Location: Top
  Spacing: Planted in patches
  Propagation Method: Seed
Probable Origin: New World

44. PARSLEY   Petroselinum sativum Hoffm.; Perijil
Varieties: 1
Uses:
  Food: Condiment
  Other: Economic, ritual
Planting Data:
  Incidence: Occasional**
  Time: Year-round
  Location: Top
  Spacing: Planted in patches
  Propagation Method: Seed
Probable Origin: Old World

45. ANISE   Pimpinella anisum L.; Anís
Varieties: 1
Uses:
  Food: Condiment
  Other: Medicinal, ritual, economic
Planting Data:
  Incidence: Rare***
  Time: Year-round
  Location: Top
  Spacing: Isolated specimens
  Propagation Method: Seed
Probable Origin: Old World

Table A.1 (cont'd)
Cultivated Tablón Crops

---

46. THYME   Thymus _vulgaris_ L.; Tomillo
Varieties: 1
Uses:
  Food: Condiment
  Other: Economic, ritual
Planting Data:
  Incidence: Occasional**
  Time: Year-round
  Location: Top
  Spacing: Planted in patches
  Propagation Method: Seed
Probable Origin: Old World

47. DILL   Anethum _graveolens_ L.; Eneldo
Varieties: 1
Uses:
  Food: Condiment
  Other: Medicinal, ritual, economic
Planting Data:
  Incidence: Rare***
  Time: Year-round
  Location: Top
  Spacing: Planted in patches
  Propagation Method: Seed
Probable Origin: Old World

48. MINT   Mentha _viridis_ L.; Hierba Buena
Varieties: 1
Uses:
  Food: Condiment
  Other: Medicinal, ritual, economic
Planting Data:
  Incidence: Rare***
  Time: Year-round
  Location: Top
  Spacing: Planted in patches
  Propagation Method: Seed
Probable Origin: Old World

MEDICINALS

49. LEMON GRASS   Cymbopogon _citratus_ (DC.) Stapf; Té de limon; limonad
Varieties: 1
Uses:
  Food: Snack food
  Other: Medicinal, economic, ritual, social
Planting Data:
  Incidence: Rare***
  Time: Year-round
  Location: Peripheral (not in tablón proper), top
  Spacing: Isolated specimens
  Propagation Method: Seed
Probable Origin: Old World

50. BORAGE   Borago _officinalis_ L.; Borage; p'o'rax
Varieties: 1
Uses:
  Food: No food value
  Other: Medicinal, economic, ritual
Planting Data:
  Incidence: Rare***
  Time: Year-round
  Location: Margins (on top), side, top
  Spacing: Isolated patches
  Propagation Method: Seed
Probable Origin: Old World

Table A.1 (cont'd)
Cultivated Tablón Crops

51. ----   Gomphrena globosa; Amor Seca
Varieties: 1
Uses:
  Food: No food value
  Other: Medicinal, economic, ritual
Planting Data:
  Incidence: Occasional**
  Time: Year-round
  Location: Top
  Spacing: Isolated specimens
  Propagation Method: Seed
Probable Origin: ?

52. BASIL   Ocimum basilicum L.; Albahaca
Varieties: 1
Uses:
  Food: Condiment
  Other: Medicinal, economic, ritual
Planting Data:
  Incidence: Rare***
  Time: Year-round
  Location: Top
  Spacing: Isolated specimens
  Propagation Method: Seed
Probable Origin: Old World

53. VERVAIN   Verbena spp.; Verbena; wer'wen
Varieties: 1
Uses:
  Food: No food value
  Other: Medicinal, economic, ritual
Planting Data:
  Incidence: Rare***
  Time: Year-round
  Location: Top
  Spacing: Isolated specimens
  Propagation Method: Seed
Probable Origin: Old World

54. RUE   Ruta muralis; Ruda; rur
Varieties: 1
Uses:
  Food: No food value
  Other: Medicinal, economic, ritual
Planting Data:
  Incidence: Rare***
  Time: Year-round
  Location: Top
  Spacing: Isolated specimens
  Propagation Method: Seed
Probable Origin: Old World

55. POINSETTIA   Euphorbia pulcherrima Willd.; Flor de Pascua
Varieties:
Uses:
  Food: No food value
  Other: Medicinal, economic, ritual
Planting Data:
  Incidence: Rare***
  Time: Year-round
  Location: Top
  Spacing: Isolated specimens
  Propagation Method: Seed
Probable Origin:

Table A.1 (cont'd)
Cultivated Tablón Crops

56. CHAMOMILE	_Matricaria chamomilla_ _L._; Manzanilla
Varieties:	
Uses:	
Food:	No food value
Other:	Medicinal, economic, ritual
Planting Data:	
Incidence:	Rare***
Time:	
Location:	
Spacing:	
Propagation Method:	
Probable Origin:	Old World

FRUIT TREES

57. ORANGE	_Citrus sinensis_ (_L._) Osg.; Naranja; a'lan
Varieties:	3
Uses:	
Food:	Snack food
Other:	Economic, ritual, social
Planting Data:	
Incidence:	Rare***
Time:	Year-round
Location:	Top
Spacing:	Isolated specimens
Propagation Method:	Seed/transplant
Probable Origin:	Old World

58. LIME	_Citrus limon_ (_L._) Burm. f.; Limón; lim
Varieties:	1
Uses:	
Food:	Snack food
Other:	Medicinal, economic, ritual, social
Planting Data:	
Incidence:	Rare***
Time:	Year-round
Location:	Top
Spacing:	Isolated specimens
Propagation Method:	Seed/transplant
Probable Origin:	Old World

59. LEMON-LIME CITRUS	_aurantiifolia_ (Christm.) Swingle var. limetta; Lima-limón; li'mon)
Varieties:	1
Uses:	
Food:	Snack food
Other:	Medicinal, economic, ritual, social
Planting Data:	
Incidence:	Rare***
Time:	Year-round
Location:	Top
Spacing:	Isolated specimens
Propagation Method:	Seed/transplant
Probable Origin:	Old World

60. PEACH	_Prunus persica_ (_L._) Botsch; Durazano; tu'ra?s
Varieties:	2
Uses:	
Food:	Snack food
Other:	Economic, ritual, social
Planting Data:	
Incidence:	Rare***
Time:	Year-round
Location:	Top
Spacing:	Isolated specimens
Propagation Method:	Seed/transplant
Probable Origin:	Old World

Table A.1 (cont'd)
Cultivated Tablón Crops

61. MANGO   Mangifera indica L.; Mango; mank
Varieties: 1
Uses:
  Food: Snack food
  Other: Economic, ritual, social
Planting Data:
  Incidence: Occasional**
  Time: Year-round
  Location: Top
  Spacing: Isolated specimens
  Propagation Method: Seed
Probable Origin: Old World

62. AVOCADO   Persea americana Mill.; Aguacate; ox
Varieties: 4
Uses:
  Food: Side dish
  Other: Economic, ritual, social
Planting Data:
  Incidence: Rare***
  Time: Year-round
  Location: Top
  Spacing: Isolated specimens
  Propagation Method: Seed/transplant
Probable Origin: New World

63. "SPANISH PLUM"   Spondias spp.; Jocote; q'u'mum
Varieties: 9
Uses:
  Food: Snack food
  Other: Economic, ritual, social
Planting Data:
  Incidence: Rare***
  Time: Year-round
  Location: Top
  Spacing: Isolated specimens
  Propagation Method: Seed/transplant
Probable Origin: New World

64. GUAVA   Pidium Guajava L.; Guava; i'kjeq
Varieties: 2
Uses:
  Food: Snack food
  Other: Ritual
Planting Data:
  Incidence: Rare***
  Time: Year-round
  Location: Top
  Spacing: Isolated specimens
  Propagation Method: Seed/transplant
Probable Origin: New World

65. NANCE   Byrsonima crassifolia (L.) H. B. et K.; Nance; ta'pal
Varieties: 2
Uses:
  Food: Snack food
  Other: Economic, ritual, social
Planting Data:
  Incidence: Rare***
  Time: Year-round
  Location: Top
  Spacing: Isolated specimens
  Propagation Method: Seed/transplant
Probable Origin: New World

Table A.1 (cont'd)
Cultivated Tablón Crops

---

66. SAPOTE, GREEN   Calocaysum viride Pitt.; Ingerto; tu'lul
Varieties: 3
Uses:
  Food: Snack food
  Other: Economic, ritual, social
Planting Data:
  Incidence: Rare***
  Time: Year-round
  Location: Top
  Spacing: Isolated specimens
  Propagation Method: Seed/transplant
Probable Origin: New World

67. SAPOTE, WHITE   Casimiroa cervantesia; Zapote blanco
Varieties: 3
Uses:
  Food: Snack food
  Other: Economic, ritual, social
Planting Data:
  Incidence: Rare***
  Time: Year-round
  Location: Top
  Spacing: Isolated specimens
  Propagation Method: Seed/transplant
Probable Origin: New World

68. MANTASANO   Casimiroa Sapota Oerst.; Matasano; axa'teʃl
Varieties: 3
Uses:
  Food: Snack food
  Other: Economic, ritual, social
Planting Data:
  Incidence: Rare***
  Time: Year-round
  Location: Top, house yards
  Spacing: Isolated specimens
  Propagation Method: Seed/transplant
Probable Origin: New World

69. ANONA   Annona spp.; Anona; p k
Varieties: 3
Uses:
  Food: Snack food
  Other: Medicinal, economic, ritual, social
Planting Data:
  Incidence: Rare***
  Time: Year-round
  Location: Top, house yards
  Spacing: Isolated specimens
  Propagation Method: Seed/transplant
Probable Origin: New World

70. PAPAYA   Carica papaya L.; Papaya
Varieties: 2
Uses:
  Food: Snack food
  Other: Medicinal, economic, ritual, social
Planting Data:
  Incidence: Occasional**
  Time: Year-round
  Location: Top, house yards
  Spacing: Isolated specimens
  Propagation Method: Seed/transplant
Probable Origin: New World

Table A.1 (cont'd)
Cultivated Tablón Crops

PLANTS USED FOR FIBER, FUEL AND/OR CONSTRUCTION

71. YUCCA   Yucca elephantipes Regel; Hizote; par'kij
Varieties: 1
Uses:
  Food: No food value
  Other: Technological, economic
Planting Data:
  Incidence: Occasional**
  Time: Year-round
  Location: Margins (on top), side, top, peripheral (not tablón proper)
  Spacing: Isolated specimens
  Propagation Method:
Probable Origin: New World

72. RIVER CANE   Gynerium saccharoides Humb. et Bonpl.; Caña Veral; ax
Varieties: 1
Uses:
  Food: No food value
  Other: Technological, economic
Planting Data:
  Incidence: Occasional**
  Time: Year-round
  Location: Peripheral (not in tablón proper), top
  Spacing: Planted in patches
  Propagation Method:
Probable Origin: New World

73. SAUCE   Salix chilensis; Sauce; si'kap'
Varieties: 1
Uses:
  Food: No food value
  Other: Technological, economic
Planting Data:
  Incidence: Rare***
  Time: Year-round
  Location: Peripheral (not in tablón proper), side
  Spacing: Isolated specimens
  Propagation Method: Seed/transplant
Probable Origin: New World

74. ALDER   Alnus acuminata; llamo; la'ma
Varieties: 1
Uses:
  Food: No food value
  Other: Technological, economic
Planting Data:
  Incidence: Rare***
  Time: Year-round
  Location: Top, seed beds and transplant, house yards, peripherals
  Spacing: 8 inches, isolated specimens
  Propagation Method: Seed/transplant
Probable Origin:

75. PRICKLY PEAR   Opuntia spp.; Tuna
Varieties: 1
Uses:
  Food: Snack food
  Other:
Planting Data:
  Incidence: Occasional**
  Time: Year-round
  Location: Peripheral (not in tablón proper)
  Spacing: Isolated specimens
  Propagation Method:
Probable Origin: New World

Table A.1 (cont'd)
Cultivated Tablón Crops

ORNAMENTALS

76. MARIGOLD *Tagetes patula* L.; Flor de Muerto
Varieties:
Uses:
   Food: No food value
   Other: Ritual, social, economic
Planting Data:
   Incidence: Occasional**
   Time: Year-round
   Location: Top
   Spacing: Isolated specimens, planted in patches
   Propagation Method: Seed
Probable Origin: New World

77. DAISY *Callistephus chinensis* (*L.*) Nees; Margarita
Varieties:
Uses:
   Food: No food value
   Other: Ritual, economic, social
Planting Data:
   Incidence: Occasional**
   Time: Year-round
   Location: Top
   Spacing: Planted in patches
   Propagation Method: Seed
Probable Origin: New World

78. GREVILEA *Grevillea robusta* A. Cunn.; Gravilea
Varieties:
Uses:
   Food: No food value
   Other: Ritual, economic, social
Planting Data:
   Incidence: Occasional**
   Time: Year-round
   Location: Top
   Spacing: Planted in patches, isolated specimens
   Propagation Method: Seed
Probable Origin: New World

79. TOUCH-ME-NOT *Impatiens balsamica*; Flor de China
Varieties:
Uses:
   Food: No food value
   Other: Ritual, economic, social
Planting Data:
   Incidence: Occasional**
   Time: Year-round
   Location: Top
   Spacing: Planted in patches, isolated specimens
   Propagation Method: Seed
Probable Origin: Old World

80. EASTER LILY *Lilium candidum*; Azucena
Varieties: 2
Uses:
   Food: No food value
   Other: Ritual, economic, social
Planting Data:
   Incidence: Occasional**
   Time: Year-round
   Location: Top
   Spacing: Planted in patches, isolated specimens
   Propagation Method: Seed
Probable Origin: Old World

Table A.1 (cont'd)
Cultivated Tablón Crops

81. CANNA   Canna spp. L.; Cucuyûs
Varieties: 5
Uses:
  Food: No food value
  Other: Ritual, economic, social
Planting Data:
  Incidence: Occasional**
  Time: Year-round
  Location: Top
  Spacing: Planted in patches, isolated specimens
  Propagation Method: Seed
Probable Origin: New World

82. PANSY   Viola tricolor L.; Pensamiento
Varieties:
Uses:
  Food: No food value
  Other: Ritual, economic, social
Planting Data:
  Incidence: Occasional**
  Time: Year-round
  Location: Top
  Spacing: Planted in patches, isolated specimens
  Propagation Method: Seed
Probable Origin: New World

*Greater than 35 percent.

**Between 5 and 35 percent.

***Less than 5 percent.

NOTE: Items listed in "Other Uses" are listed in order of importance.

Appendix B

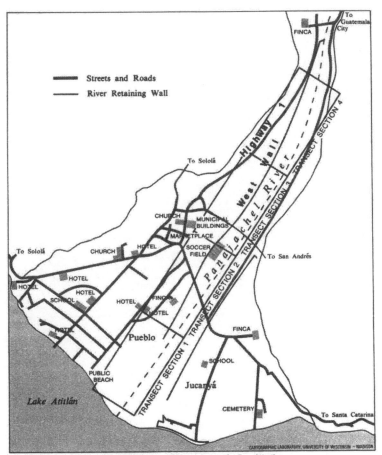

Panajachel: The Area of the Study

Detailed maps corresponding to the transect sections above appear on the following pages.

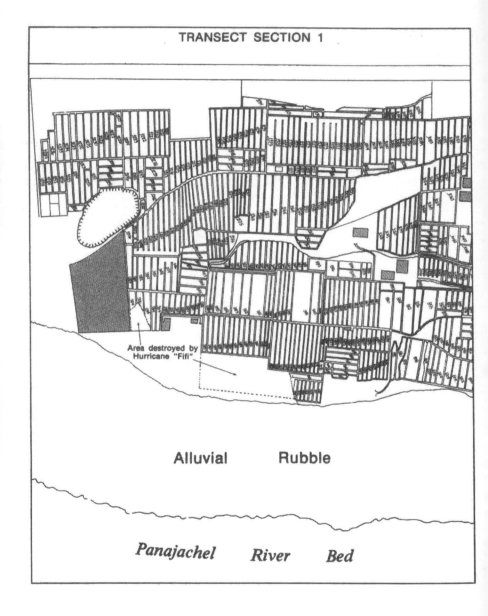

TRANSECT SECTION 1

Area destroyed by
Hurricane "Fifi"

Alluvial   Rubble

Panajachel   River   Bed

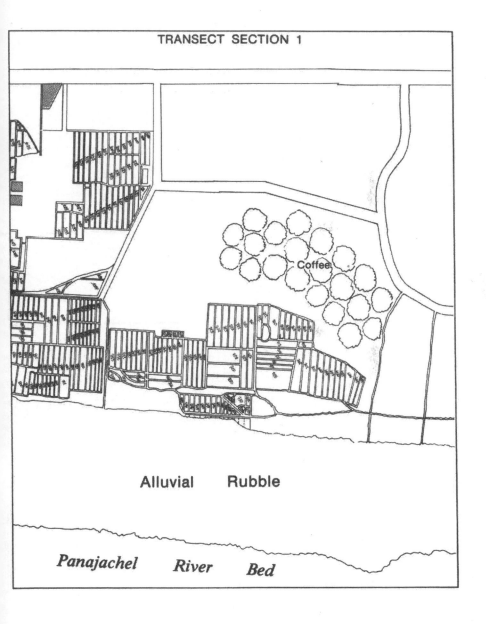

TRANSECT SECTION 1

Coffee

Alluvial   Rubble

Panajachel   River   Bed

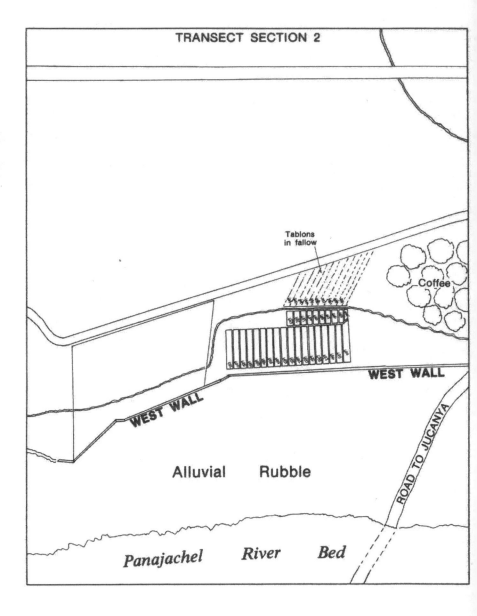

TRANSECT SECTION 2

Tablons
in fallow

Coffee

WEST WALL

WEST WALL

Alluvial   Rubble

Panajachel   River   Bed

ROAD TO JUCANYA

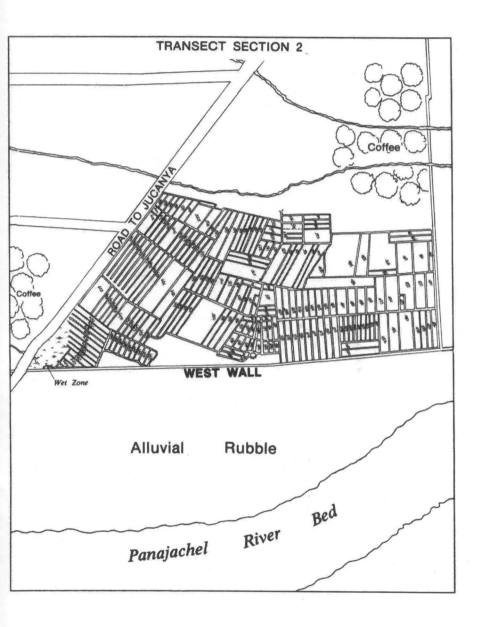

TRANSECT SECTION 2

Coffee

ROAD TO JUCANYA

Coffee

Wet Zone

WEST WALL

Alluvial   Rubble

Panajachel   River   Bed

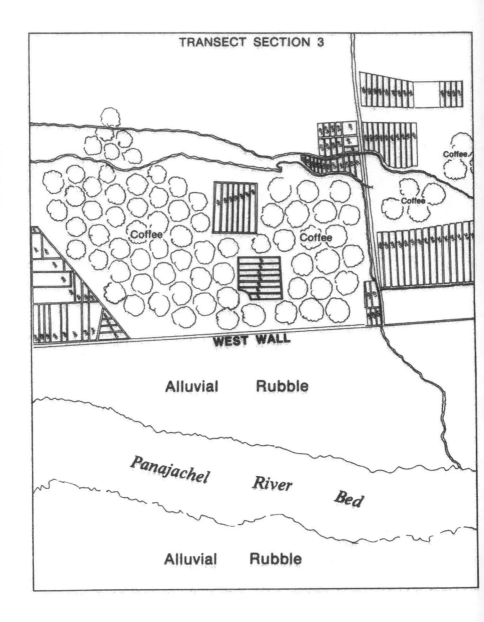

TRANSECT SECTION 3

Coffee

Coffee

Coffee

Coffee

Coffee

WEST WALL

Alluvial Rubble

Panajachel River Bed

Alluvial Rubble

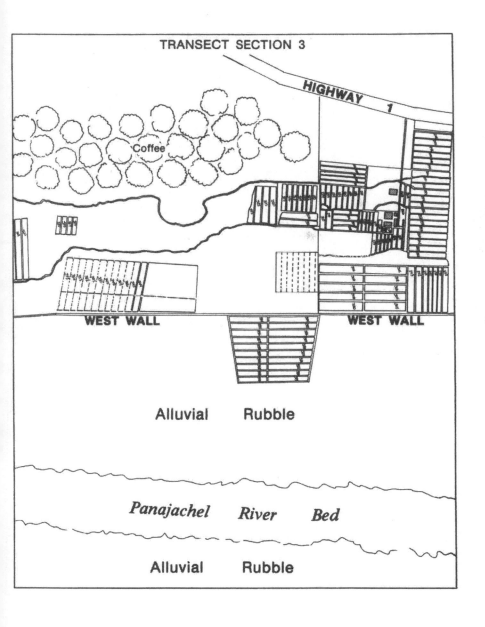

TRANSECT SECTION 3

HIGHWAY 1

Coffee

WEST WALL

WEST WALL

Alluvial   Rubble

Panajachel   River   Bed

Alluvial   Rubble

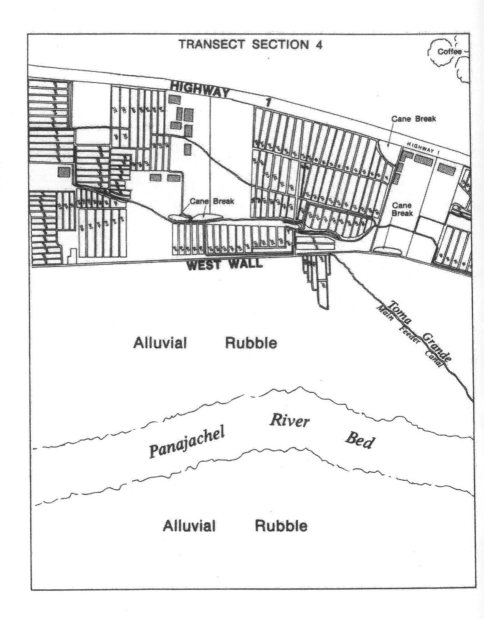

TRANSECT SECTION 4

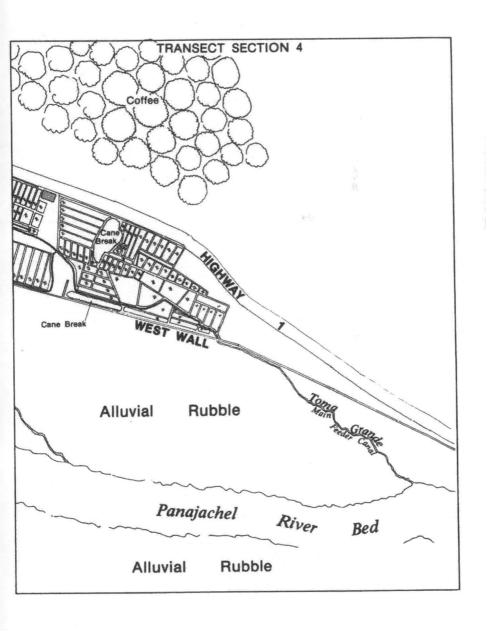

Index

Printed and bound by CPI Group (UK) Ltd, Croydon, CR0 4YY

23/10/2024

01778241-0011